Emergent Directions in Engineering Education

Volume 1

The book series *Emergent Directions in Engineering Education* provides a dedicated forum to explore innovative and transformative developments, highlighting emerging visions and future-oriented directions, in engineering education. Driven by advances in digital technologies, interdisciplinary practices, and increasing attention to social, ethical, and sustainability considerations, engineering education is evolving rapidly. This series captures and synthesises these dynamic changes through scholarly contributions that bridge research, theory, and practice.

The series invites conceptual and philosophical perspectives, rigorous theoretical discussions, research syntheses, and empirical studies, across all themes of relevance to engineering education, which aim to foster a deeper understanding and advancing of the field. By presenting scholarship in an accessible and integrated manner, this series encourages dialogue among engineering educators, researchers, and practitioners across diverse sub-disciplines.

Designed for engineering education researchers at all career stages, as well as faculty, instructors, curriculum designers, academic administrators, industry professionals, and policymakers, the series uniquely addresses a critical gap by providing a platform for comprehensive exploration and interdisciplinary engagement. It encourages submissions from a diverse authorship, ensuring openness to new ideas and perspectives from across the global engineering education community. Book proposals for this series may be submitted to the Publishing Editor: Claudia Acuna E-mail: Claudia.Acuna@Springer.com

Ardion Beldad • Pascal Wilhelm
Jasper Homminga
Editors

Designing an Interdisciplinary Engineering Education for Complex Real-World Challenges

Springer

Editors
Ardion Beldad
University of Twente
Enschede, The Netherlands

Pascal Wilhelm
Saxion University of Applied Sciences
Deventer, The Netherlands

Jasper Homminga
University of Twente
Enschede, The Netherlands

ISSN 3091-3136 ISSN 3091-3144 (electronic)
Emergent Directions in Engineering Education
ISBN 978-3-032-20222-2 ISBN 978-3-032-20223-9 (eBook)
https://doi.org/10.1007/978-3-032-20223-9

This Springer imprint is published by the registered company Springer Nature Switzerland AG
The registered company address is: Gewerbestrasse 11, 6330 Cham, Switzerland

Preface

A Liberal Arts Engineering Education for Modern-Day Grand Challenges

It has been three years since the World Health Organisation declared the end of the COVID-19 pandemic as a global health emergency in May 2023 (United Nations, 2023). As the world has returned to normalcy, people's memories of the psychological, social, and economic havoc wrought by the pandemic have more likely waned. Nonetheless, one compelling lesson to be learned from how the world coped with the pandemic is the importance of integrated and carefully coordinated solutions from a multiplicity of expertise to address an array of complex challenges. For instance, as Moradian et al. (2020) noted, at the height of the pandemic, medical scientists, responsible for infection rate monitoring and pandemic preparedness, were hardly sufficient on their own. The pandemic, the authors add, also required biological scientists to understand how pathogens interact with their hosts, engineers to develop effective protective equipment to curtail infection risks, and social scientists to analyse the implications of government-imposed measures for citizens' mental and emotional well-being.

Grand challenges confronting modern societies abound—from climate change to geopolitical conflicts, and from the vicious cycle of poverty and inequality to the steady escalation of physical and mental health crises. Continuous technological advancement and society's growing dependence on technology also instigate other grand challenges, such as disinformation and cybersecurity threats. George et al. (2016) define a grand challenge as a 'specific critical barrier(s) that, if removed, would help solve an important societal problem with a high likelihood of global impact through widespread implementation' (p. 1881). Any grand challenge is inherently complex (Ackermann, 2024), demanding solutions generated through collaboration among various social actors possessing the needed expertise and experience (Bates et al., 2022; George et al., 2016).

Solutions to grand challenges entail behavioural modifications at individual and societal levels and require technological innovations for problem-solving (George et al., 2016). Clearly, a disciplinary-siloed approach has become inadequate in addressing complex issues. Hence, it is unsurprising that, in recent years, calls for interdisciplinary collaboration from civic societies, policymakers, and the business sector have intensified. Even research agencies in the European Union and the United States stimulate interdisciplinary research projects (Pedersen, 2016).

The urgent need for engineers who can tackle complex problems by engaging with other disciplinary experts (Van den Beemt et al., 2020) prompts an expanding number of academic institutions to offer interdisciplinary educational programmes, such as the Massachusetts Institute of Technology's undergraduate programme in urban science, the bachelor's and master's programme in technical medicine at the University of Twente (the Netherlands), the master's programme in energy science and technology at ETH Zurich (Switzerland), and the master's programme in interdisciplinary engineering at Texas A&M University (USA).

PhD programmes with a strong interdisciplinary character are also offered in several universities, such as those specialising in global studies (University of Geneva, Switzerland) and interdisciplinary humanities (King's College London, UK). James Jacob (2014) opines that this considerable increase in interdisciplinary educational programmes is a response to the changing demands of the labour market, especially as employers value graduates who can function effectively in teams and possess core disciplinary competencies.

Designing education centred on interdisciplinary problem-solving can be approached in two ways: first, by having students from different disciplines collaborate to address a challenge; and second, by requiring students who are proficient in the languages and methods of multiple disciplines to tackle a challenge (Klaassen, 2018). For the second approach, adopting a liberal arts educational philosophy is particularly fitting, as it exposes students to a range of academic disciplines before they pursue disciplinary specialisation in the latter part of their education (Hudson & Hinman, 2017; Nussbaum, 2002).

The open nature of a liberal arts programme provides fertile ground for interdisciplinarity to flourish. Relatively unrestricted in its course offerings, a liberal arts programme provides students with ample opportunities to develop a myriad of disciplinary competencies essential for interdisciplinary problem-solving. Additionally, the small-scale character of a liberal arts programme (Astin, 1999) nurtures a sense of community and connection indispensable for collaborative efforts (Han et al., 2022), especially when tackling grand challenges. With liberal arts students eventually specialising in different fields, collaboration on educational activities benefits from the diverse expertise, perspectives, and mindsets of team members, resulting in outcomes that could essentially be characterised as interdisciplinary.

In recent years, liberal arts education has experienced a surge in popularity in Europe, with programmes in the Netherlands, Germany, and Switzerland. At the time of writing, the Netherlands has nine university colleges (UCs) offering a bachelor's programme in liberal arts, with almost all being part of larger Dutch research universities and are publicly funded (Van der Wende, 2011). Van der Wende (2011) attributes the growing popularity of the liberal arts programme in Europe, in general, and in the Netherlands, in particular, to the demand for a broader bachelor's programme that counters issues related to early specialisation or overspecialisation at the undergraduate level and to the need to differentiate from the rigid, monodisciplinary orientation of European university programmes. Furthermore, the author claims that renewed enthusiasm for liberal arts education in the Netherlands is fuelled by the race to attract global talent, a growing market for intellectually challenging programmes, and opportunities for students to specialise in the sciences within a liberal arts framework.

With engineering curricula becoming more interdisciplinary (Feng et al., 2023; Van den Beemt et al., 2020), it seems natural to infuse them with a liberal arts character. Systematic efforts to integrate the liberal arts in engineering have been increasingly pursued in various forms. Hitt et al. (2023) refer to these as programmatic integrations. One approach is to provide engineering students with opportunities to do a minor in the arts or humanities. Designing educational activities requiring engineering and non-engineering students to collaborate in solving grand challenges is another approach. Integration is also pursued by introducing liberal arts topics (e.g., identities, virtue ethics, social and environmental justice) throughout an engineering curriculum (Hitt et al., 2023).

Equipping students with interdisciplinary competencies through courses across the sciences is one thing; providing them with opportunities to apply integrated knowledge is another. Problem-based learning (PBL), reported to be the most frequently used pedagogical approach in interdisciplinary engineering settings (Van den Beemt et al., 2020), is instrumental in promoting knowledge integration. A prerequisite for implementing PBL is to have real-world challenges right at the core of the learning process (Kolmos, 2009), with students actively involved in knowledge construction throughout the problem-solving exercise (Stentoft, 2017). PBL's student-centred philosophy requires self-directed and critical students to effectively perform certain tasks (e.g., selection of materials needed to understand a challenge) (Loyens et al., 2008).

The vision of an interdisciplinary engineering education utilising the PBL approach and entrenched in a liberal arts educational philosophy with a strong emphasis on self-directed learning resulted in an educational experiment at the University of Twente, located in the eastern part of the Netherlands and one of the four Dutch technical universities. In 2013, the university launched the University College Twente, which offers a three-year bachelor's programme in Technology, Liberal Arts, and Sciences, officially abbreviated as ATLAS. A distinct feature of the programme is its emphasis on educating 'new engineers' who embody five core values, namely self-directed learning, interdisciplinarity, personal development, trailblazing, and community. This emphasis fuels the vision of educating interdisciplinary engineers who are both intellectually competent and socially and ethically responsible.

An engineering curriculum rooted in a liberal arts educational philosophy somehow defies what Van der Wende (2011) describes as the prevailing practice of steering students to specialise upon entering a European university system. Calls for broadening the engineering curriculum have been intensifying in recent years (e.g.,

Conlon, 2013; Duderstadt, 2010). The educational experiment at the University of Twente is a response to those calls and serves as a starting point for this edited volume.

Structure of this Edited Volume

This book commences with a chapter on the societal value of inter- and transdisciplinary (ITD) education by Boor, de Greef, and Wenting. The authors reflect on the forces undergirding the ongoing transition from monodisciplinary education to interdisciplinary and transdisciplinary pedagogical approaches. They argue that the societal value of ITD is hinged on its role of educating students who can navigate complex societal challenges, its instrumental function in driving innovation, its mission of preparing future-ready professionals, and its potential to fulfil students' desire to be responsible members of society.

Boor et al. describe the bachelor-level Change Making Project at the University of Amsterdam, which enables students to experience tackling grand societal challenges by collaborating with professionals across various disciplines. The project also introduces students to the Transitional Cycle methodology, consisting of four phases: imagine (e.g., developing a vision for societal change), connect (e.g., identifying enablers and hindrances to vision realisation), act (e.g., execution of a short-term action), and assess (e.g., reflection on the created intervention). The authors claim that despite the societal value of ITD education, its implementation faces challenges that must be addressed both at the curricular and institutional levels.

Liberal arts programmes also face a myriad of challenges, but their social value cannot be downplayed, as echoed in Chap. 2 by Kumar, Nair, Upadhyay, and Anand. The authors contend that liberal arts education is pivotal in preparing students for the evolving landscape of the engineering profession. They refute the view that liberal arts education has reached obsolescence. Instead, they underscore the important roles such an education plays in nurturing students' interdisciplinary thinking, critical thinking, problem-solving, and communication skills, strengthening their adaptability, and cultivating their propensity to engage in lifelong learning.

An important point advanced in this chapter is the need for liberal arts institutions to adapt their curricula to the changing demands of the workforce, in particular, and the economy, in general. The pervasiveness of artificial intelligence, for instance, requires future professionals not only proficient in data science but also adept at addressing philosophical questions arising from people's interaction with the technology. According to the authors, despite the benefits of liberal arts education, its practical value (e.g., the employability of its graduates) is constantly being questioned.

In Chap. 3, Wilhelm, Beldad, Homminga, Garcia Chapeton, and Pereira Pessoa describe the design of an interdisciplinary engineering programme founded on a liberal arts educational philosophy. The three-year bachelor's programme—Technology, Liberal Arts, and Sciences (ATLAS)—fully embraces the view that the natural sciences, the social sciences, and mathematics are equally important in equipping engineers with the required competencies to address modern-day grand challenges. According to the authors, the liberal arts educational philosophy, with its emphasis on exposing students to a range of disciplines and its use of student-centred educational approaches, corresponds to an interdisciplinary engineering curriculum's goal of educating well-rounded engineers competent in interdisciplinary problem-solving. The educational programme's adoption of project-based learning ensures that students are acquiring and developing competencies indispensable for collaborative problem-solving.

Wilhelm et al. note that the three-year interdisciplinary engineering education is predicated on a set of values (e.g., interdisciplinarity, self-directed learning) that steer the programme's final learning outcomes (FLOs), primarily driven by the mission to educate a new breed of engineers who are competent in the different disciplinary lenses for understanding (natural science, social science, and mathematics) and the processes of creation (through research and design), have strong communication and collaboration skills, and possess well-developed academic attitude. The emphasis on these four competencies in the programme's FLOs requires broadly formulated learning goals at the semester level. Such a broad formulation fosters students' autonomy in selecting their educational activities based on their learning needs, learning goals, and academic interests.

Wilhelm et al. further describe how the six semesters of the three-year bachelor's programme are designed to facilitate students' realisation of the FLOs, even with a unique educational curriculum for each student. At the core of the first four semesters is a project on a particular grand challenge that students must collaboratively tackle. This approach directly implements project-based learning (PBL), which, as several authors claim (e.g., MacLeod & Van der Veen, 2020; Nielsen et al., 2010; Van der Beemt et al., 2020), is an important teaching method for cultivating students' interdisciplinary competencies.

Towards the end of the third chapter, the authors discuss the structure of two semester projects representing different levels of complexity and scope. The semester project in the first semester focuses on designing physical solutions for a specific target group, while the third semester project introduces students to the process of solving wicked problems across the three phases of mapping and modelling, mitigation strategy creation, and integration of developed mitigation strategies.

One of the defining features of the ATLAS programme, as described in the third chapter, is its emphasis on self-directed learning (SDL). In Chap. 4, Wilhelm reviews the benefits students derive from an SDL-based education. As students assume control of the learning process in an SDL-based education, they become more effective and motivated learners, and are very likely to pursue lifelong learning. This chapter primarily focuses on the ways the three year-bachelor's programme reconciles the tension between implementing SDL and delivering quality education.

Wilhelm asserts that while students have the autonomy to design their own curriculum, the educational programme also employs a monitoring mechanism to ensure that curricular choices made by students will enable them to reach the programme's final learning outcomes. This mechanism, referred to as the SDL cycle, involves three phases that require students to plan, evaluate, and reflect throughout the semester. These phases include writing a personal development plan (PDP), doing a mid-term evaluation (MTE), and performing a self-evaluation and reflection (SER).

With the social sciences securing a prominent position in an interdisciplinary engineering curriculum, the design and implementation of social science courses warrant serious attention. Chap. 5 by Beldad outlines the requirements for designing mandatory social science courses within the aforementioned curriculum. The author argues that the role the social sciences play in educating a new breed of engineers cannot be discounted anymore, especially as they will be confronted with grand challenges demanding not only technical but also social solutions.

Beldad argues that a critical challenge for a social science course in an interdisciplinary engineering programme is striking a balance between adequately introducing students to the discipline's language and methods and exposing them to theories and concepts that matter when developing solutions to complex real-world problems. The broad character of an interdisciplinary engineering education requires academic staff involved in mandatory courses, including those in the social sciences, to painstakingly design learning activities that would expand students' disciplinary knowledge base and develop their academic and professional skills. According to the author, it is important to focus on relevant analytical frameworks, perspectives, and concepts when developing courses within a semester dedicated to a particular grand societal challenge. Another important consideration is to emphasise perspectives and theories broadly applicable across the social science discipline, and those that can explain the roles of various social actors in implementing engineering solutions and the individual values and social forces shaping the interactions between actors and solutions.

In Chap. 6, Kittou underscores the indispensability of mathematics in interdisciplinary engineering education, specifically the role it plays in cultivating students' modelling skills. Mathematical modelling techniques, the author claims, help interdisciplinary engineering students to analyse real-world problems systematically. The importance of modelling prompted the design of a mandatory course entitled Introduction to Mathematical Modelling, consisting of three components, namely preparation (e.g., familiarisation with a topic), active engagement (e.g., application of concepts to real-world cases), and reflection and refinement (e.g., reflection on feedback received).

The author emphasises the utility of the Double Diamond Design Thinking Process, introduced by the Design Council in 2003 (Design Council, n.d.), as a foundation for the pedagogical approach of implementing activities requiring scientific inquiry in a mathematics course, promoting interdisciplinary thinking through a combination of skills in physics, calculus, and data analysis, and encouraging both individual and team efforts of problem-solving. Kittou also reflects on the curricular and organisational challenges of designing a mathematics curriculum for an interdisciplinary engineering programme, and argues that curricular flexibility and organisational commitment to the education's learning outcomes can help resolve those challenges.

The selection of knowledge and skills to emphasise in mandatory physics courses within an interdisciplinary engineering programme is also a challenge confronting physics teachers. Chapter 7 by Van der Hoef, Wildeman, and Homminga introduces a concept-based approach for developing physics courses that relate to the grand challenges driving semester projects and those that are sufficiently interesting for interdisciplinary engineering students with a more social science orientation.

Using this concept-based approach, the authors identified four interrelated macro-concepts that serve as the basis for designing mandatory physics courses: systems thinking, mathematisation of phenomena, origin and hierarchy of laws, and energy. Just like in the mathematics course described in the previous chapter, mandatory physics courses in an interdisciplinary engineering programme also emphasise the importance of the modelling process as a component of solving real-world problems. The four macro-concepts, Van der Hoef et al. argue, can be directly related to the modelling process. In the second part of Chap. 7, the authors describe how two mandatory courses—Systems & Transport Phenomena and Particles & Collective Phenomena—are designed and implemented.

In Chap. 8, Dijksterhuis discusses the implementation of the Personal Pursuit in a three-year bachelor's programme in interdisciplinary engineering education. This Personal Pursuit resembles a personal project, which, according to Little (2020), is potentially useful in giving meaning, structure, and direction to an individual's life. When done in an educational environment, such a personal project, Little claims, provides students with opportunities to cultivate creativity and experience novel ways of learning. In this chapter, Dijksterhuis posits that the Personal Pursuit is an embodiment of self-directed learning, as it invites students to transform a personal passion or interest (e.g., wildlife photography, learning a foreign language, producing a documentary) into an educational project and requires them to design their own learning by formulating learning objectives, specifying educational activities, and identifying the final product that will be assessed.

The eighth chapter elaborates on the organisation of the Personal Pursuit and its three key components—self-directed learning, learning outcomes, and guidance (through tutoring). Dijksterhuis accentuates the role of students' Personal Pursuit in empowering them to be effective self-directed learners.

Beldad and Miedema (2026) argue that an interdisciplinary engineer must not only have disciplinary knowledge and skills (e.g., research, collaboration, communication) but should also maintain an academic mindset, which they define as a 'constellation of attitudes towards and beliefs in the value of self-awareness to realise opportunities for academic and personal growth, in the relevance of critical and analytical thinking to understand themselves and to comprehend phenomena and realities, and in the importance of behaving in an ethically acceptable manner' (p. 197). This emphasis on consistently fostering an academic mindset in an interdisciplinary engineering programme is a response to calls from professional engineering associations (e.g., National Academy of Engineering, Royal Academy of Engineering) to educate engineers who 'understand the social, ethical, cultural, political, and environmental contexts of their work' (p. Hitt et al., 2023, p. 457).

This focus on embedding values alongside academic skills development in an interdisciplinary engineering programme is the focus of Chap. 9 by Kizito. The author examines the ways the aforementioned programme uses the learning line approach to accentuate the centrality of academic, ethical, and social mindsets in the education of a new breed of engineers. Inspired by cases of successful values integration in engineering curricula (e.g., Delft University of Technology, Ollin College), the interdisciplinary engineering programme's Academic Approach and Attitude (AAA) learning line (alongside several learning lines) provides a basis for a segment of its final learning outcomes.

The author introduces an AAA learning line framework that highlights the importance of self-awareness for both intellectual and character development. While the implementation of the learning line throughout the three-year bachelor's programme benefits efforts of educating a complete engineering professional, implementation-related challenges are also copious. Kizito recommends specific actions to deal with the challenges.

Research-based evidence for the positive effects of self-directed learning on students' attitude, character, behaviour, and skills development abounds. Nonetheless, the role of experience with an SDL-based education in shaping the attitude and behaviour of alumni in their professional and personal lives remains underexplored. Chapter 10 by Kampouroglou, Beldad, and Wilhelm focuses on how years of experience with SDL impacted the professional functioning of graduates of an interdisciplinary engineering programme, and on the role of such an experience in those graduates' personal lives.

Based on data obtained through in-depth interviews with 26 graduates across 9 cohorts (from 2016 to 2024), the authors found that SDL, commonly viewed by participants as a way of thinking and as a tool, has facilitated reflection skills development, increased individual sense of autonomy, and enhanced their self-confidence—three important outcomes that are helping them function well in their professional and personal lives. The positive effects of SDL, nonetheless, are also countered by certain issues, such as graduates becoming overcritical of themselves and struggling to deal with failures, especially when being self-directed has often resulted in positive results. Moreover, even after graduating from an SDL-based bachelor's programme, the majority of interviewees still maintain a positive attitude towards SDL, primarily due to the benefits of being self-directed learners.

The last chapter of this volume critically reflects on the challenges confronting an interdisciplinary engineering programme founded on a liberal arts educational philosophy. Beldad, Wilhelm, and Homminga contend that developing, organising, and implementing a bachelor's programme in interdisciplinary engineering, with its emphasis on self-directed learning and problem-based learning, is associated with several curricular and organisational challenges. For instance, at the curricular level, the tension between maintaining a broad curriculum throughout the three-year programme and having one that leads to disciplinary specialisation persists. Ensuring a strong connection between courses and the project within a semester is another hurdle.

Additionally, there are also organisational challenges that merit attention. For instance, an interdisciplinary education requires an interdisciplinary teaching staff, but the monodisciplinary orientation of each member might pose difficulties resulting from marked differences in disciplinary languages and power dynamics. However, as this last chapter demonstrates, specific actions can still be pursued to mitigate those challenges.

Enschede, The Netherlands Ardion Beldad

References

Ackermann, F. (2024). Managing grand challenges: Extending the scope of problem structuring methods and behavioural operational research. *European Journal of Operational Research, 319*(2), 373–383. https://doi.org/10.1016/j.ejor.2024.01.024

Astin, A. W. (1999). How the liberal arts college affects students. *Daedalus, 128*(1), 77–100. http://www.jstor.org/stable/20027539

Bates, R., Brenner, B., Schmid, E., Steiner, G., & Vogel, S. (2022). Towards meta competences in higher education for tackling complex real-world problems – A cross-disciplinary review. *International Journal of Sustainability in Higher Education, 23*(8), 290–308. https://doi.org/10.1108/IJSHE-06-2021-0243

Beldad, A. D., & Miedema, H. A. T. (2026). Introducing a framework for designing an interdisciplinary engineering curriculum: Educating new engineers for complex sociotechnical challenges. *European Journal of Engineering Education, 51*(1), 192–209. https://doi.org/10.1080/01972243.2011.583802

Conlon, E. (2013). Broadening engineering education: Bringing the community in: Commentary on "Social responsibility in French engineering education: A historical and sociological analysis". *Science and Engineering Ethics*, 19(4), 1589–1594. https://doi.org/10.1007/s11948-013-9476-x

Design Council. (n.d.). *The Double Diamond.* Design Council. https://www.designcouncil.org.uk/our-resources/the-double-diamond/

Duderstadt, J. J. (2010). Engineering for a changing world: A roadmap to the future of American engineering practice, research, and education. In D. Grasso & M. B. Burkins (Eds.), *Holistic engineering education: Beyond technology* (pp. 17–35). Springer New York.

Feng, X., Ylirisku, S., Kähkönen, E., Niemi, H., & Hölttä-Otto, K. (2023). Multidisciplinary education through faculty members' conceptualisations of and experiences in engineering education. *European Journal of Engineering Education, 48*(4), 707–723. https://doi.org/10.1080/03043797.2023.2185126

George, G., Howard-Grenville, J., Joshi, A., & Tihanyi, L. (2016). Understanding and tackling societal grand challenges through management research. *Academy of Management Journal, 59*(6), 1880–1895. https://doi.org/10.5465/amj.2016.4007

Han, J., Jiang, Y., Mentzer, N., & Kelley, T. (2022). The role of sense of community and motivation in the collaborative learning: An examination of the first-year design course. *International Journal of Technology and Design Education, 32*(3), 1837–1852. https://doi.org/10.1007/s10798-021-09658-6

Hitt, S.J., Banzaert, A., & Pierrakos, O. (2023). Educating the whole engineer by integrating engineering and the liberal arts. In A. Johri (Ed.), *International handbook of engineering education research* (pp. 457–476). Taylor & Francis.

Hudson, P. F., & Hinman, S. E. (2017). The integration of geography in a curriculum focused to internationalization: An interdisciplinary liberal arts perspective from the Netherlands. *Journal of Geography in Higher Education, 41*(4), 549–561. https://doi.org/10.1080/03098265.2017.1337089

James Jacob, W. (2015). Interdisciplinary trends in higher education. *Palgrave Communications, 1*(1), 1–5. https://doi.org/10.1057/palcomms.2015.1

Klaassen, R. G. (2018). Interdisciplinary education: A case study. *European Journal of Engineering Education, 43*(6), 842–859. https://doi.org/10.1080/03043797.2018.1442417

Kolmos, A. (2009). Problem-based and project-based learning. In O. Skovsmose, P. Valero, & O.R. Christensen (Eds.), *University science and mathematics education in transition* (pp. 261–280). Springer.

Little, B. R. (2020). How are you doing, really? Personal project pursuit and human flourishing. *Canadian Psychology/Psychologie canadienne, 61*(2), 140–150. https://psycnet.apa.org/doi/10.1037/cap0000213

Loyens, S. M., Magda, J., & Rikers, R. M. (2008). Self-directed learning in problem-based learning and its relationships with self-regulated learning. *Educational Psychology Review, 20*(4), 411–427. https://doi.org/10.1007/s10648-008-9082-7

MacLeod, M., & Van der Veen, J. T. (2020). Scaffolding interdisciplinary project-based learning: A case study. *European Journal of Engineering Education, 45*(3), 363–377. https://doi.org/10.1080/03043797.2019.1646210

Moradian, N., Ochs, H. D., Sedikies, C., Hamblin, M. R., Camargo Jr, C. A., Martinez, J. A., … & Rezaei, N. (2020). The urgent need for integrated science to fight COVID-19 pandemic and beyond. *Journal of Translational Medicine, 18*(1), 1–7. https://doi.org/10.1186/s12967-020-02364-2

Nielsen, J. D., Du, X. Y., & Kolmos, A. (2010). Innovative application of a new PBL model to interdisciplinary and intercultural projects. *International Journal of Electrical Engineering Education, 47*(2), 174–188. https://doi.org/10.7227/IJEEE.47.2.7

Nussbaum, M. (2002). Education for citizenship in an era of global connection. *Studies in Philosophy and Education, 21*, 289–303. https://doi-org.ezproxy2.utwente.nl/10.1023/A:1019837105053

Pedersen, D. B. (2016). Integrating social sciences and humanities in interdisciplinary research. *Palgrave Communications, 2*(1), 1–7. https://doi.org/10.1057/palcomms.2016.36

Stentoft, D. (2017). From saying to doing interdisciplinary learning: Is problem-based learning the answer? *Active Learning in Higher Education, 18*(1), 51–61. https://doi.org/10.1177/1469787417693510

United Nations. (2023, May 5). *WHO chief declares end to COVID-19 as a global health emergency*. UN News. https://news.un.org/en/story/2023/05/1136367

Van den Beemt, A., MacLeod, M., Van der Veen, J., Van de Ven, A., Van Baalen, S., Klaassen, R., & Boon, M. (2020). Interdisciplinary engineering education: A review of vision, teaching, and support. *Journal of Engineering Education, 109*(3), 508–555. https://doi.org/10.1002/jee.20347

Van der Wende, M. (2011). The emergence of liberal arts and sciences education in Europe: A comparative perspective. *Higher Education Policy, 24*(2), 233–253. https://doi.org/10.1057/hep.2011.3

Contents

About the Contributors

Nivedita Anand, MSc, is the head of branding at Thapar Institute of Engineering & Technology, where she leads communication, marketing, and event branding. A graduate of the Erasmus Mundus Master's in Research and Innovation in Higher Education (MARIHE), she has studied in Austria, Finland, India, Japan, and the Netherlands. Her academic and professional interests lie at the intersection of innovation, pedagogy, and organisational change.

Ardion Beldad, PhD, is an associate professor at the University of Twente, the Netherlands. He belongs to a core team of academic staff that founded the University College Twente in 2012. While helping to establish the new interdisciplinary engineering programme, he also taught at the communication science department of the UT. As a social scientist, he conducts research on trust creation, crisis management, information privacy, prosocial behaviour and ethical consumption, and technology adoption.

Ilja Boor, PhD, is head of the Teaching & Learning Centre Science at the University of Amsterdam, the Netherlands and programme lead of the Visible Learning Trajectories Program. She leads large-scale curriculum innovation and teacher team professionalisation and focuses on genAI and affective learning. She is a co-author of Meaningful Assessment in Interdisciplinary Education.

Fokko Jan Dijksterhuis, PhD, is an associate professor in the history of science and technology at the University of Twente and the Vrije Universiteit Amsterdam. Throughout his academic career, his focus on teaching has been on interdisciplinarity, educating engineering students in societal aspects and humanities students in science and technology. He has been involved in numerous educational innovations (e.g., honours programmes at UT & VU, developing the University College Twente).

Gustavo Adolfo García Chapeton, PhD, is a senior researcher in cybertechnology and artificial intelligence within the research group Technologies for Criminal Investigation at the Academy of Life Sciences, Engineering, and Design, Saxion University of Applied Sciences, the Netherlands. His work focuses on developing methods to enhance criminal investigation processes. He coordinated the semester project on wicked problems at the University College Twente.

Linda de Greef, MSc, is innovation lead at the Institute for Interdisciplinary Studies, University of Amsterdam, the Netherlands. She pioneers inter- and transdisciplinary education, designs transformative curricula, facilitates lecturer professionalisation and is actively involved in national and international educational innovation networks. She is a co-author of Meaningful Assessment in Interdisciplinary Education.

Martin van der Hoef, PhD, is an associate professor at the Faculty of Science and Technology, the University of Twente, with a research background in numerical simulation of multiphase flow. In 2012, he switched his focus to teaching when he joined the core team of teachers that founded the University College Twente. Next to this, he is a teacher in the natural sciences for the Chemical Science and Engineering and the Advanced Technology programmes at the University of Twente.

Jasper Homminga, PhD, is a biomechanical engineer with a focus on designing and delivering education. He has designed education in fields ranging from medicine to engineering. He has co-designed curricula, such as Biomedical Engineering and Technology and Liberal Arts & Sciences. He has been the programme director for two programmes: Technical Medicine and Technology and Liberal Arts & Sciences.

Thiseas Kampouroglou, BSc, graduated from the University College Twente, University of Twente, the Netherlands, with a degree in technology, liberal arts, and sciences. His research interests are focused on educational and sustainability philosophy, primarily from the lenses of feminism and critical theory.

Georgia Kittou, PhD, is a lecturer at the University College Twente, University of Twente, the Netherlands. She teaches mathematics in a project-based, interdisciplinary engineering programme. Her research interests focus on designing strategies that enhance student motivation and foster deeper engagement using narratives, storytelling, and gamified learning experiences. Her scholarly background is in mathematical physics and theoretical cosmology.

Rita Ndagire Kizito, PhD, is a learning architect and scholar with 30+ years in higher education leadership, open and distance learning, and teaching across Africa and Europe. Her scientific activities focus on the learning sciences and culturally grounded design for STEM/interdisciplinary education. She pioneered the Academic Approach & Attitude Framework for embedding self-directed learning and social responsibility, cultivating ethical new engineers.

Vinay Kumar, PhD, is a professor and dean at the Thapar School of Liberal Arts & Sciences, Patiala, India. His research ranges from artificial intelligence, computer vision and cloud-based medical imaging, and technology-driven pedagogy. With his team, he is building a technology-driven liberal arts and sciences school while championing innovative, student-centred pedagogy.

Padmakumar Nair, PhD, is the Vice-Chancellor of Thapar Institute of Engineering & Technology. An accomplished scholar in the areas of leadership development, sustainability, and nanomaterials, he specialises in strategising for sustainability and social & commercial Entrepreneurship. He holds two PhDs—one in nano-structured materials from the University of Twente, Netherlands, and another in chemical engineering from the University of Tokyo, Japan.

Marcus Vinicius Pereira Pessoa, PhD, is an associate professor at the University of Twente's Systems Engineering and Multidisciplinary Design Chair and the head of educational innovation at the UT's Centre of Expertise in Learning and Teaching. His research interests are systems modelling and architecture using model-based systems engineering, programme management, and education as a system.

Rahul Upadhyay, PhD, is an associate professor at the Thapar School of Liberal Arts & Sciences, Patiala, India. His work lies at the intersection of AI, neuroscience, and signal processing, with a focus on brain–computer interfaces, biomedical applications, and cognitive technologies. His research explores how AI and interdisciplinary approaches can advance neuromarketing, enrich traditional knowledge systems, and drive socially impactful innovation.

Lucy Wenting, MSc, is the director of the Institute for Interdisciplinary Studies at the University of Amsterdam, the Netherlands. She specialises in interdisciplinary and transdisciplinary teaching and learning, strategy development, coaching, and process facilitation, with a proven track record in educational leadership at the institutional level. She is also a trainer of the UvA Educational Leadership Programme.

Sander Wildeman, PhD, is an experimental physicist and lecturer in the natural sciences at the University of Twente, the Netherlands. He developed expertise in creating courses in which students understand the natural world and engineering problems by applying a conceptual lens to real-world phenomena. In his course designs, hands-on experimentation and physical-mathematical modelling are tightly intertwined.

Pascal Wilhelm, PhD, is a developmental and educational psychologist. He was part of the team that designed ATLAS, an interdisciplinary bachelor's engineering programme at the University of Twente, the Netherlands, based on the concept of self-directed learning. He currently works as a teacher and researcher at the Academy of Applied Psychology and Human Resource Management of Saxion University of Applied Sciences in the Netherlands.

Chapter 1
The Societal Value of Inter- and Transdisciplinary Education

Ilja Boor, Linda de Greef, and Lucy Wenting

1.1 Introduction

In recent years, universities worldwide have increasingly shifted from mono-disciplinary programmes towards inter- and transdisciplinary (ITD) forms of education. This shift reflects a growing recognition that the pressing challenges of the twenty-first century (e.g., climate change, digital transformation, and sustainable health systems) are not only technically complex but also value-laden, involving societal values and normative considerations. These challenges cut across disciplinary boundaries, demand multi-actor responses, and require new forms of integrative learning. As a result, higher education is challenged to prepare graduates who can navigate, integrate, and act within contexts marked by complexity, uncertainty, and normative plurality.

Against this backdrop, this chapter examines the societal relevance of ITD education as a structural necessity for universities seeking to remain connected and relevant to society. While disciplinary expertise remains indispensable, curricula organised solely along disciplinary lines cultivate depth in specific domains. ITD education responds to these for the so-called wicked problems (Rittel & Webber, 1973) by equipping students to integrate perspectives, collaborate across boundaries, and evaluate the broader societal implications of their work.

The shift towards ITD education is also shaped by evolving societal expectations. Universities are increasingly seen as actors with a public mandate for

I. Boor (✉)
Teaching & Learning Centre Science, University of Amsterdam,
Amsterdam, The Netherlands
e-mail: p.k.i.boor@uva.nl

L. de Greef · L. Wenting
Institute for Interdisciplinary Studies, University of Amsterdam, Amsterdam, The Netherlands

A. Beldad et al. (eds.), *Designing an Interdisciplinary Engineering Education for Complex Real-World Challenges*, Emergent Directions in Engineering Education 1, https://doi.org/10.1007/978-3-032-20223-9_1

responsibility to contribute to sustainable development, democratic resilience, and inclusive innovation (OECD, 2020). This expectation is evident in policy frameworks worldwide. The European Union's agenda for higher education emphasises interdisciplinarity as a key condition for tackling grand societal challenges (European Commission, 2020). UNESCO (2023) underscores the importance of transdisciplinary learning for achieving the Sustainable Development Goals across regions. The US National Academies (2020) have called for increased interdisciplinarity in sustainability education, combined with diversity, equity, and inclusion. These calls point to an expanded conception of the university's role: not only producing knowledge but also cultivating professionals who can apply knowledge responsibly in societal contexts. ITD education embodies this expansion by making explicit the links between knowledge, skills, values, and action.

1.2 Unravelling ITD Education

Understanding the societal value of ITD education requires first clarifying the distinct modes of cross-disciplinary engagement, with multidisciplinary as the most basic form, denoting the side-by-side use of insights from different disciplines without systematic integration (Klein, 2008).

Interdisciplinarity goes further by integrating methods, concepts, and theories to generate new understandings unattainable by a single discipline alone. It involves addressing topics too broad or complex for one discipline, drawing on multiple fields to construct a more comprehensive perspective (Klein & Newell, 1997). Boix Mansilla et al. (2000) describe interdisciplinarity as the capacity to integrate knowledge and modes of thinking from two or more disciplines or established areas of expertise to achieve cognitive advancement, such as explaining a phenomenon, solving a problem, or creating a product that would have been impossible or unlikely within a single discipline.

Transdisciplinarity expands this scope by incorporating extra-academic knowledge sources such as entrepreneurs, policymakers, community members, patient organisations, and practitioners. This is increasingly important because current and future societal challenges cannot be addressed effectively by scientific or technological solutions isolated from social and practical contexts (Bernstein, 2015; Klein et al., 2017). As Roux et al. (2017), Pohl et al. (2017), and the OECD (2020) argue, challenges that affect society at large require solutions that integrate both scientific and societal dimensions. In transdisciplinary education, academic and extra-academic knowledge are thus combined to generate new insights collaboratively between scholars and stakeholders (Bernstein, 2015; Pohl et al., 2017).

Across these modes, the hallmark of ITD education is integration. One driver of integration is to develop a comprehensive understanding of complex questions. It is rooted in intellectual curiosity and the ambition to explore knowledge at the intersection of disciplines (Repko et al., 2019). Advancement in the fundamental understanding is the guiding principle behind interdisciplinary education. The second driver is more societally oriented, as it focuses on integration as a means to address and solve urgent and complex real-world challenges, such as those related to health, food, water, energy, or globalisation (Hadorn et al., 2008). This form of integration often defines transdisciplinary education. In practice, the boundaries between these orientations are not strict. Curiosity-driven integration can also benefit from transdisciplinary approaches, especially when extra-academic perspectives help reframe or deepen academic understanding. Likewise, addressing societal challenges often requires rigorous interdisciplinary integration to synthesise insights from multiple academic fields.

Integration may involve the integration of various research elements-roughly referring to different phases of the research project, such as theories and concepts, research methods and design, and data and results. For example, distinct concepts and theories from different disciplines may need to be integrated to ensure the formulation of a comprehensive research problem and determine specific research questions. Additionally, different tools and techniques can be combined to better address the complexity of the question and yield more robust results. Regarding the results of a research project, different sets of data or results produced by using different methods can complement each other and be integrated, leading to a deeper understanding (Keestra et al., 2022; Repko & Szostak, 2025). Often, transdisciplinary projects tend to produce practical interventions or instruments. The integration of extra-academic knowledge, such as indigenous knowledge, personal experience, emotions, intuition, implicit and explicit societal rules, and belief systems, is then also incorporated (Scholz & Steiner, 2015).

1.3 The Four Why's

What defines the societal value of ITD education? This value can be articulated through four interconnected rationales, each shedding light on a distinct yet mutually reinforcing dimension: (1) navigating complex societal challenges, (2) driving innovation, (3) preparing future-ready professionals, and (4) responding to students' desire for societal relevance and responsibility. Each rationale is explored through mechanisms, implications for education, and illustrative examples. Together, they provide a framework for understanding why universities should invest in ITD education.

1.3.1 Navigating Complex Societal Challenges

Complex challenges lie at the heart of ITD studies. Such challenges are shaped by multiple interacting factors and demand collaboration across fields of study and an approach that integrates diverse perspectives. ITD studies are shaped by dual drivers: the intellectual pursuit of new insights at disciplinary intersections, and the societal need to confront urgent, complex challenges that transcend single fields (Hadorn et al., 2008; Repko et al., 2019).

An example of such a challenge is the growing tension between technological developments, surveillance, privacy, and public safety. Advancements in artificial intelligence, data collection, and predictive analytics have created powerful tools for monitoring and preventing threats. At the same time, they raise serious concerns about privacy, discrimination, misuse of data, and the erosion of civil liberties. This dilemma cannot be addressed by technology or law alone. It requires input from computer science, ethics, law, sociology, political science, and the voices of citizens affected by these systems. An ITD program or course provides students with the ability to navigate these tensions, approach issues from multiple perspectives, critically evaluate the societal consequences, and learn to co-develop responsible and sustainable solutions.

ITD curricula should be designed to provide structured opportunities for students to practice these skills, encouraging systematic and recurring engagement throughout the program. Beyond exposure to multiple disciplines, these curricula open pathways for making connections, finding common ground, identifying patterns across fields, and critically engaging with different perspectives while navigating the uncertainties inherent in ITD.

Lecturers play an essential role by guiding students through conflicting perspectives and modelling how partial insights can be brought into dialogue. By doing so, they help students see how diverse viewpoints can be integrated into more comprehensive understandings. Collaborative learning settings are particularly valuable in this regard. Confronting unfamiliar perspectives deepens reflection, supports knowledge exchange, and mitigates common pitfalls such as overconfidence and groupthink. Learning methods such as design thinking, placemaking, responsible futuring, and changemaking are widely used to support this process of integrating diverse disciplinary perspectives (Greef et al., in press).

1.3.2 Driving Innovation

ITD programmes equip students with the capacity to advance innovation across disciplinary and societal domains. Although innovation is frequently defined as the generation of new solutions, it can be more precisely understood as arising at the intersection of creativity and changemaking (Greef et al., in press). Creativity refers to the capacity to generate original, valuable ideas, while changemaking provides the purposeful drive to translate such ideas into meaningful impact. Only when creative possibilities are implemented in ways that create tangible value do they become innovation. The value of creativity has been underscored by the World Economic Forum (2023), which lists it among the ten most important skills for future employees. ITD education promotes creativity by training students to imagine possibilities beyond disciplinary boundaries, and it strengthens changemaking by equipping them with the skills to implement these possibilities by interventions in collaborative and impactful ways.

Innovation thrives at points of intersection: between disciplines, between academic and societal domains, and between established and emerging forms of knowledge. Such boundary crossings spark unique insights that cannot be confined within traditional boundaries. ITD education positions students in precisely these spaces of overlap and tension. By navigating them, students build the capacity to connect ideas, to experiment, and to adaptively pursue opportunities. Engineers provide an exemplary case whose role will extend beyond the technical domain. They are expected not only to design efficient systems but also to address ethical dilemmas, mediate between societal stakeholders, and align technological possibilities with sustainable futures.

ITD education thus lays the groundwork for engineers to act as responsible changemakers in society. The implication for education is that, for example, engineering curricula should not only focus on rigorous knowledge and critical thinking, but also embed opportunities for experimentation and problem solving. Lecturers play a crucial role by modelling how creative ideas can be tested, refined, and transformed into impactful innovations. Their task goes beyond delivering content. They create learning environments where uncertainty is embraced rather than avoided, where failure is treated as a source of insight, and where students are encouraged to engage with perspectives different from their own.

1.3.3 Preparing Future-Ready Professionals

Higher education carries a growing responsibility to prepare students for a world marked by uncertainty, rapid technological change, and societal transformation. Graduates will increasingly face challenges whose solutions are contested and whose outcomes are unpredictable. In such a context, professional readiness cannot be defined solely in terms of disciplinary expertise. Instead, it requires a broader set of transferable capabilities: critical thinking, collaboration, creativity, reflection, and changemaking (Greef et al., in press). ITD education offers a promising approach for cultivating these capabilities.

The OECD's Future of Education and Skills 2030 project highlights three transformative competencies for navigating uncertainty: creating new value, reconciling tensions and dilemmas, and taking responsibility (OECD, 2018). ITD education aligns strongly with this vision. By confronting students with problems that cannot be solved within disciplinary boundaries, ITD education compels them to develop adaptive mindsets. Students learn to negotiate between different epistemic traditions, to anticipate the societal consequences of technical choices, and to deliberate value trade-offs with diverse stakeholders.

Yet knowledge and skills alone are not sufficient for preparing future-ready professionals. As several scholars have argued, addressing complex societal challenges also requires the cultivation of values and attitudes (OECD, 2018; Fischman &

Gardner, 2022; Gooblar, 2019). Higher education must therefore engage students in forms of affective learning: the development of attitudes, beliefs, and emotional capacities that shape motivation, ambition, and responsible decision-making (Biesta, 2020; UNESCO, 2015). Such capacities include strengthening an inner compass and staying true to core values and purpose, while also cultivating openness, receptiveness to unfamiliar perspectives, and resilience, which entails the ability to recover from setbacks while maintaining focus and adaptability (Krathwohl et al., 1964).

Fields such as sustainability have long stressed the importance of combining cognitive and affective learning. The Inner Development Goals (IDG) framework, developed by the Ekskäret Foundation, offers inspiration for embedding these affective dimensions in ITD curricula. It outlines 23 skills, attitudes, or values across the five dimensions—Being, Thinking, Relating, Collaborating, and Acting—that complement the Sustainable Development Goals. The Institute for Interdisciplinary Studies at the University of Amsterdam, in collaboration with the alliance of Technical University Eindhoven, Wageningen University & Research, Utrecht University and University Medical Centre Utrecht, created a Toolbox for Transition Makers based on the IDG framework. It serves as an extensive resource for teachers to design intended learning outcomes, develop learning activities, and implement assessment methods in higher education.

1.3.4 Responding to Students' Desire for Relevance and Responsibility

In a comparative analysis of six European countries regarding students' perspectives on the purpose of higher education, it was found that students primarily view higher education as serving three main objectives: preparing for the job market, fostering personal growth and enrichment, and contributing to societal development and progress. Students want curricula that let them engage with socially urgent issues, learn how to make meaningful contributions to society, and experience personal growth—areas they feel are not sufficiently addressed in their current education (Brooks et al., 2021). ITD education can address this demand by structuring learning experiences around relevance and responsibility.

By embedding authentic learning experiences, students are able to connect their academic knowledge and skills to real-world challenges. One such pedagogical approach is challenge-based learning, a didactical approach in which students tackle complex and meaningful problems that demand critical thinking, collaboration, and creativity. A defining feature of challenge-based learning is its transdisciplinary character: students not only work across academic disciplines, but also collaborate closely with extra-academic stakeholders who help define the challenges, co-develop solutions, and provide feedback throughout the process.

A concrete illustration can be found in the minor Biomimicry at the University of Amsterdam, where students use Design Thinking methods to develop prototypes—often robotic—that address the needs of potential end-users. In this minor, students explore how nature builds strong and lightweight materials, forms networks, heals itself, and senses its environment, and then apply these insights to design a technologically advanced, bio-inspired prototype. For instance, they might replicate the gripping properties of a bear claw, design a honeycomb-inspired structure for strength and efficiency, or create an autonomous robot. Each team selects its own challenge, translates biological principles into a robot, and demonstrates their prototypes at a final symposium.

1.4 Case Study: The Change Making Project

To illustrate how the societal value of ITD education materialises in practice, this section presents the "Change Making Project" (worth 12 European credits equivalent to 336 hours of coursework). It is offered to second- and third-year bachelor students at the University of Amsterdam. Students from a wide range of disciplinary and cultural backgrounds participate in this course. The course aims to convey to students that a reflective, experimental attitude is key to developing the ability to cope with complex societal challenges and enable societal transitions. The course emphasises an experimental attitude, encouraging students to adopt an open-minded and critical approach when addressing real-world issues. To achieve this, the lecturers developed a co-learning environment where students collaborate with professionals from various fields. Collaboration partners are identified and confirmed before the start of the course, allowing student teams to engage immediately with relevant societal challenges. The core of the course is built around the Transition Cycle methodology, guiding students through four iterative phases: Imagine, Connect, Act, and Assess (Bertolini et al., 2024).

In the 'Imagine' phase, students define a "What if…?" question and collaboratively develop a vision for societal change. This phase encourages creative thinking through workshops that focus on things such as spoken word, theatre, or science fiction. Students are prompted to imagine alternative futures by posing questions such as: "What if… we had ownership over the data we create? What if… we stopped eating junk food? What if… there were no filter bubbles on the internet? What if… everyone had equal access to healthcare?".

After the vision has been developed, students proceed to the 'Connect' phase, where they learn about systems thinking and create system maps and stakeholder analyses. These tools help them understand why their vision is not yet a reality and identify the complexities of the systems they are addressing. In this phase, students often recognise the complexity of their chosen systems. They identify enablers and obstacles for their future vision and see how different factors and actors in the system are related to and dependent on each other. They question the status quo and ask themselves for example: 'Why do we do things the way we do?' and 'Why are streets for cars and not for people?' Then students realise all the power structures, dependencies, assumptions, cultural aspects, and regulations related to car use.

The final phases, Act and Assess, are closely interconnected. The 'Act' phase begins with a roundtable discussion featuring "changemakers" from fields such as art, science, activism, and entrepreneurship. Students then develop and execute a transition experiment, which is a short-term action through which alternative structures, cultures, and practices are explored. The students set up a monitoring plan for the intervention to reflect and learn from their intervention. An example of such an experiment is a guerrilla clean-up in the Vondelpark in Amsterdam, to make people aware of trash in public spaces and motivate them to participate in volunteering.

In the 'Assess' phase, students reflect on their interventions and the entire 'Transition Cycle' process. This reflection forms the basis of their final portfolio submission and helps students evaluate their own growth and contributions to societal challenges. Throughout the course, students submit draft products after each phase. Teachers and peers provide feedback to ensure continuous improvement. At the end of the course, students submit a final group portfolio covering the four phases, along with an individual reflection on their learning journey. The portfolio consists of the vision and a creative product (for example, a poem or drawing) of the 'Imagine' phase, a system- and stakeholder map and analysis for the 'connect' phase, and an action plan and impact evaluation for the 'Act' and 'Assess' phases. The main benefit of the portfolio is that the assessment is cumulative: several smaller assignments, which are provided with feedback, add up to the final portfolio and grade. Students appreciate how continuous feedback helps them to improve the final product. The final grade covers mostly the learning process of the students - and not only the end product - in order to leave room for experimentation and failure.

The Change Making Project resonates with today's generation of students, who are already confronted with global challenges that will continue to shape their future. Alongside a knowledge base, the course emphasises personal growth and self-reflection as essential foundations that enable students to translate concern into meaningful change.

1.5 What Is at Stake for Higher Education?

In recent years, universities worldwide have increasingly shifted from monodisciplinary programmes towards ITD forms of education, acknowledging that ITD education creates societal value by enabling graduates to understand and address complex, value-laden challenges that cannot be solved within the confines of a single discipline. ITD education confronts these challenges directly, catalysing innovation, preparing future-ready professionals, and responding to students' demand for relevance and responsibility.

This transformation requires more than curricular reform. It calls for learning environments that deliberately foster collaboration and societal engagement, and for lecturers who act not only as transmitters of knowledge but as facilitators of transformative learning processes. For students, ITD education creates relevance, agency, and readiness for uncertainty.

For education leaders, researchers, lecturers, and developers alike, this presents both a strategic and a pedagogical challenge: how do we create space for this alternative mode of education? How do we enable genuine collaboration across departments, with societal partners, and with students as co-creators of learning?

The future of ITD in higher education will demand vision, courage, and institutional imagination. Not because it is easy, but because it is essential. Even in uncertain times, universities must invest in and reward these practices. In doing so, they will not only cultivate resilient graduates but also strengthen their credibility as institutions that can meaningfully address pressing societal challenges. Taking ITD education seriously is, therefore, not an option but a condition for remaining relevant in today's world.

References

Bernstein, J. H. (2015). Transdisciplinarity: A review of its origins, development, and current issues. *Journal of Research Practice, 11*(1), 1–20.

Bertolini, L., Gerritsen, D., & Sol, K. (2024). Supporting students to become agents of change: Introducing and evaluating the transition cycle approach to teaching transformative skills. *Futures, 164*, 1–14. https://doi.org/10.1016/j.futures.2024.103459

Biesta, G. (2020). Risking ourselves in education: Qualification, socialization, and subjectification revisited. *Educational Theory, 70*(1), 89–104. https://doi.org/10.1111/edth.12411

Boix Mansilla, V., Miller, W. C., & Gardner, H. (2000). *On disciplinary lenses and interdisciplinary work*. Interdisciplinary Studies Project, Project Zero, Harvard Graduate School of Education.

Brooks, R., Gupta, A., Jayadeva, S., & Abrahams, J. (2021). Students' views about the purpose of higher education: A comparative analysis of six European countries. *Higher Education Research & Development, 40*(7), 1375–1388. https://doi.org/10.1080/07294360.2020.1830039

de Greef, L., Boor, I, & Wenting, L.U. (in press). *Designing inter- and transdisciplinary education: A handbook for university teachers*. Taylor & Francis.

European Commission, Directorate-General for Education, Youth, Sport and Culture. (2020). *Prospective report on the future of transnational collaboration in European higher education*. Publications Office of the European Union. https://data.europa.eu/doi/10.2766/098140

Fischman, W., & Gardner, H. (2022). *The real world of college: What higher education is and what it can be*. MIT Press.

Gooblar, D. (2019). *The missing course: Everything they never taught you about college teaching*. Harvard University Press.

Hadorn, G. H., Hoffmann-Riem, H., Biber-Klemm, S., Grossenbacher-Mansuy, W., Joye, D., Pohl, C., & Zemp, E. (2008). *Handbook of transdisciplinary research*. Springer.

Keestra, M., Uilhoorn, A., & Zandveld, J. (2022). *An introduction to interdisciplinary research* (2nd rev. ed.). Amsterdam University Press.

Klein, J. T. (2008). Evaluation of interdisciplinary and transdisciplinary research: A literature review. *American Journal of Preventive Medicine, 35*, S116–S123. https://doi.org/10.1016/j.amepre.2008.05.010

Klein, J. T., & Newell, W. H. (1997). Advancing interdisciplinary studies. In J. G. Gaff, J. L. Ratcliff, & Associates (Eds.), *Handbook of the undergraduate curriculum: A comprehensive guide to purposes, structures, practices, and change* (pp. 393–415). Jossey-Bass.

Klein, J. T., Häberli, R., Scholz, R. W., Welti, M., & Grossenbacher-Mansuy, W. (2017). Interdisciplinarity: A critical assessment. *Minerva, 55*(4), 379–395.

Krathwohl, D. R., Bloom, B. S., & Masia, B. B. (1964). *Taxonomy of educational objectives: The classification of educational goals, Handbook II: Affective domain*. David McKay Company.

National Academies of Sciences, Engineering, and Medicine. (2020). *Strengthening sustainability programs and curricula at the undergraduate and graduate levels*. The National Academies Press. https://doi.org/10.17226/25821

Organisation for Economic Co-operation and Development (OECD). (2018). *The future of education and skills 2030: The future we want*. OECD.

Organisation for Economic Co-operation and Development (OECD). (2020). *Addressing societal challenges using transdisciplinary research.* OECD Science, Technology and Industry Policy Papers, No. 8. OECD. https://doi.org/10.1787/0ca0ca45-en

Pohl, C., Truffer, B., & Hirsch Hadorn, G. (2017). Addressing wicked problems through transdisciplinary research. In R. Frodeman, J. T. Klein, & R. C. D. S. Pacheco (Eds.), *The Oxford handbook of interdisciplinarity* (2nd ed., pp. 319–331). Oxford University Press.

Repko, A. F., & Szostak, R. (2025). *Interdisciplinary research: Process and theory* (5th ed.). Sage.

Repko, A. F., Szostak, R., & Buchberger, M. P. (2019). *Introduction to interdisciplinary studies* (2nd ed.). Sage.

Rittel, H. W. J., & Webber, M. M. (1973). Dilemmas in a general theory of planning. *Policy Sciences, 4*(2), 155–169.

Roux, D. J., Nel, J. L., Cundill, G., O'Farrell, P., & Fabricius, C. (2017). Transdisciplinary research for systemic change: Who to learn with, what to learn about and how to learn. *Sustainability Science, 12*, 711–726. https://doi.org/10.1007/s11625-017-0446-0

Scholz, R. W., & Steiner, G. (2015). The real type and ideal type of transdisciplinary processes: Part I—Theoretical foundations. *Sustainability Science, 10*, 527–544. https://doi.org/10.1007/s11625-015-0326-4

UNESCO. (2015). *Global citizenship education: Topics and learning objectives.* UNESCO. Retrieved July 2, 2025, from https://unesdoc.unesco.org/ark:/48223/pf0000233600

UNESCO. (2023). *Regional report for Asia-Pacific: Education starts early, Progress, challenges and opportunities*. UNESCO Office Bangkok and Regional Bureau for Education in Asia and the Pacific. Retrieved July 2, 2025, from https://www.unesco.org/en/articles/asia-pacific-education-2030-sdg-4-midterm-review

World Economic Forum. (2023). *What are the top 10 job skills for the future?* https://www.weforum.org/agenda/2023/05/future-of-jobs-report-2023-skills/

Chapter 2
Reimagining the Future of Work through the Lens of a Liberal Arts Education

Vinay Kumar, Padmakumar Nair, Rahul Upadhyay, and Nivedita Anand

2.1 Introduction

2.1.1 The Evolving Landscape of Work

The global economic landscape is currently undergoing a multifaceted transformation, particularly in the areas of artificial intelligence (AI), big data, and societal expectations. These powerful forces are not simply altering existing industries. They are fundamentally reshaping the very nature of work, redefining traditional job roles, and imposing novel demands on the global workforce.

The World Economic Forum (WEF) has consistently underscored this paradigm shift, highlighting the emergence of critical new skill sets essential for navigating what it terms the Fourth Industrial Revolution. Their reports suggest that reliance on traditional vocational training may no longer be adequate to prepare people for such a dynamic and often unpredictable future (World Economic Forum, n.d.). The transition from tasks that are routine and easily automatable to those demanding problem-solving, nuanced critical thinking, and creativity necessitates a comprehensive re-evaluation of our established educational frameworks.

V. Kumar (✉) · P. Nair · R. Upadhyay · N. Anand
Centre of Excellence (Policy, Pedagogy and Purpose), Thapar School of Liberal Arts & Sciences, Thapar Institute of Engineering & Technology, Patiala, India
e-mail: vinay.kumar@thapar.edu

A. Beldad et al. (eds.), *Designing an Interdisciplinary Engineering Education for Complex Real-World Challenges*, Emergent Directions in Engineering Education 1, https://doi.org/10.1007/978-3-032-20223-9_2

2.1.2 The Common Perception

Within this rapidly evolving professional environment, persistent and often vocal scepticism surrounds the utility of a liberal arts education (Sunstein, 1993; Zakaria, 2015). A common narrative suggests that such an education, traditionally rooted in the humanities, social sciences, and foundational sciences, is increasingly obsolete and impractical. Critics frequently contend that liberal arts degrees do not impart specific job-ready skills essential for immediate employment in a technology-driven world. This viewpoint often undervalues the intellectual capacities cultivated by these disciplines, contributing to the prevalent misconception that liberal arts graduates are inherently ill-equipped for the rigours and demands of the contemporary workforce.

This chapter posits a counterargument. Far from being obsolete, a liberal arts education is uniquely and powerfully positioned to furnish individuals with the indispensable, future-proof skills required in the technology-powered future of work. Liberal arts programmes prepare students for the fast-changing world by nurturing interdisciplinary thinking, critical analysis, clear and empathetic communication, and the ability to adapt with resilience—exactly the skills we need to face complex challenges and strengthen what makes us the most human.

This chapter will explore how the lasting and often underestimated value of a liberal arts education provides students with a robust intellectual framework to navigate the complexities of the twenty-first-century workforce. Additionally, to address the specific challenges and misconceptions facing liberal arts programmes regarding their practicality, we will illustrate how modern curricula—specifically through the lens of a four-component programme structure—are actively organised to bridge the gap between intellectual breadth and professional applicability. This demonstrates its extensive yet overlooked impact on innovation, ethical leadership, and broader societal well-being.

2.2 Core Tenets of Liberal Arts Education and their Enduring Value

The true value of a liberal arts education lies in its foundations. It is built not to provide narrow job training, but to nurture all-around intellectual growth and holistic development. These core principles are increasingly being recognised as absolutely critical for navigating the complexities and uncertainties that define the future of work.

2.2.1 Interdisciplinary Thinking and Application

Liberal arts education, by its very design, often facilitates and makes interdisciplinary thinking easier due to its foundational approach. It actively encourages students to forge connections and synthesise knowledge across what might initially appear to be disparate fields—ranging from insights of history and philosophy to the intricate structures of literature, the societal observations of sociology, and the rigorous methodologies of the sciences. This integrated approach contrasts traditional skill-based, often rigidly siloed, academic structures, preparing individuals to address problems that defy simplistic, single-discipline solutions. For example, dealing with difficult colleagues or wicked workplace problems (Han et al., 2023).

The cultivated ability to integrate seemingly wider ideas is of great interest in a world where wicked problems—from the existential threat of climate change to the rapid spread of global pandemics—demand interdisciplinary, collaborative, and comprehensive approaches. Such intellectual agility positions liberal arts graduates to drive innovation and develop creative solutions within the dynamic and often unpredictable nature of contemporary professional settings.

The practical application of interdisciplinary thinking through integrated courses and programmes is particularly potent in the liberal arts. For example, a course on conflict enables students to gain a holistic understanding of its multifaceted nature by incorporating various perspectives and philosophies from diverse fields such as sociology, cultural studies, theology, environmental science, computer science, and economics. This approach mirrors the real-world problems faced in the workplace, which rarely fit neatly into single academic disciplines.

2.2.2 Critical Thinking and Problem-Solving

Both liberal arts and engineering teach critical thinking, but in different ways. Liberal arts students work on complex, open-ended problems that do not have clear answers. They learn to define problems, interpret different types of evidence, and weigh ethical considerations. Engineering students focus on solving well-defined problems by creating models, working within limits, and testing whether their solutions actually work.

Liberal arts graduates are competent in figuring out what problems need solving and why they matter. Engineers excel at making sure solutions are practical, safe, and effective. The best workplaces combine both approaches—teams that can both identify the right problems and solve them well.

Thus, students of liberal arts are not simply taught what to think, but rather how to think. They are meticulously trained to question underlying assumptions, to carefully evaluate evidence, and to construct coherent, well-reasoned arguments. This process involves developing a capacity for independent thought, an ability to discern subtle biases, and the intellectual fortitude to engage deeply with complex ideas from a multitude of viewpoints. Such skills are far from being mere academic ideas; they are directly transferable to real-world problem-solving scenarios. In an age defined by information overload and the constant threat of misinformation, the capacity to critically analyse data and generate effective, evidence-based solutions remains an indispensable professional asset. Indeed, the World Economic Forum's reports consistently and unequivocally list critical thinking and complex problem-solving among the most highly sought-after skills for the future workforce (Lewis, 2018).

2.2.3 Communication and Empathy

Liberal arts programmes place a strong emphasis on developing superior communication skills, encompassing both the written and oral modalities. Students learn not only to articulate complex ideas with clarity and precision but also to do so persuasively and concisely—a competency that is absolutely vital for effective collaboration, impactful leadership, and meaningful client engagement. In the world of GenAI, these skills might look frivolous, but they are actually more important than ever. While AI can produce text quickly, it cannot analyse what different audiences need to hear, make good judgments about context, or think through ethical issues. In workplaces that use AI, people still need to check if AI-generated content is accurate and appropriate, adjust messages for different groups of people, and take responsibility for what gets communicated.

The key difference is between creating text and communicating strategically: AI handles the writing mechanics, but humans provide the thinking, relationship skills, and ethical judgment that determine whether communication actually works. Liberal arts graduates, who have developed these advanced communication abilities through their training in articulating complex ideas clearly, persuasively, and compassionately, can use AI as a tool while keeping the human insight that makes professional communication effective for collaboration, leadership, and client work.

As Edmondson et al. (2020) compellingly argue, the humanities, in particular, are instrumental in nurturing empathy and sensibility, qualities that are increasingly in demand as automation continues to absorb routine tasks. The cultivated capacity to genuinely understand and appropriately respond to the emotions and perspectives of others is fundamentally crucial for building robust teams, positive workplace cultures, and navigating the complexities of professional life. What used to be considered soft skills are rapidly ascending to the status of power skills, serving as critical differentiators of human capabilities in an increasingly AI-driven world.

2.2.4 *Adaptability and Lifelong Learning*

In an economy where job roles evolve rapidly and entire industries emerge within decades, the traditional model of training students for specific positions has become fundamentally inadequate. Liberal arts education equips students with two major components of lifelong learning: intellectual frameworks and curiosity.

It is not that skill-based graduates cannot become lifelong learners; many do. For liberal arts graduates, adaptability is more reliably cultivated when programmes are architected around sustained curiosity as a formal outcome, not only around current tools and procedures. Liberal-arts curricula embed this architecture by default: students practice open-textured inquiry, work with heterogeneous evidence, and engage in double-loop learning (questioning assumptions and goals as well as methods). These foundational reinforcements at the undergraduate level strengthen and integrate these qualities into students' intellectual character, where they continue to function as enduring features throughout their lives.

By contrast, skill-focused pathways, unless intentionally redesigned, optimise for near-term task performance and certification checklists. This yields strong near-transfer and rapid early productivity but can become brittle as tools, standards, and platforms change. When such programmes integrate meta-learning, ethical reasoning, and audience-aware communication across the curriculum, the adaptability gap closes. The difference, then, is architectural rather than essential: frameworks + curiosity can be built in any pathway, but liberal arts programmes institutionalise them, making lifelong learning an expected outcome rather than an individual accident. This, in turn, reinforces communication, compassion, and excellence-seeking as the practical disciplines that convert adaptability into responsible impact.

2.3 The Future of Work: Partnering Liberal Arts with Technical Prowess

The ongoing transformation of the global economy, driven by the relentless advancement of technologies such as artificial intelligence (AI), machine learning, big data, and advanced robotics, fundamentally alters the very fabric of the workforce. While specialised technical proficiencies undeniably retain their importance, there is an increasingly widespread consensus that the skills most resilient to automation and most pivotal for sustained success in the coming decades are precisely those traditionally nurtured and refined by a liberal arts education (Autor et al., 2003; Deming, 2017; Lassébie & Quintini, 2022; OECD, 2024; World Economic Forum, 2023).

As artificial intelligence continues its pervasive integration into various sectors, automating tasks that were once exclusively within the purview of human endeavour, the unique and irreplaceable value of the human element within the workforce becomes strikingly more pronounced. This encompasses a range of capabilities that include, but are not limited to, empathy, the capacity for complex ethical decision-making, the generation of truly novel creative ideas, and the nuanced ability to understand and adeptly navigate intricate social dynamics. These are not skills that can be readily or fully replicated by even the most advanced algorithms. For example, while AI can efficiently analyse colossal datasets to identify patterns and correlations, it fundamentally lacks the inherent human capacity for nuanced interpretation, deep contextual understanding, and the critical ethical judgment required for sensitive and responsible decision-making.

Liberal arts graduates, with their rigorous training in philosophy, ethics, history, sociology, and literature, are uniquely positioned and equipped to address the most challenging societal and ethical implications arising from emerging technologies. The future of work, therefore, is not merely about technological prowess. It is equally, if not more so, about meticulously cultivating and strategically leveraging distinctly human attributes that complement, rather than directly compete with, the capabilities of artificial intelligence.

Partnerships between liberal arts colleges and industry leaders are becoming increasingly common and robust. These collaborations facilitate invaluable mentorships, joint project initiatives, and create direct, streamlined pathways to employment for graduates. These strategic initiatives collectively demonstrate a highly proactive and adaptive approach by liberal arts institutions. They are actively working to bridge any perceived gap between academic learning and the dynamic demands of the workforce, thereby ensuring that their graduates are exceptionally well-prepared for the intricate complexities and exciting opportunities posed by the twenty-first-century job market to solve wicked problems (American Association of Colleges and Universities, 2023; Choi-Lundberg et al., 2024; Jackson & Cook, 2025).

2.4 Reimagining Curricula

In response to the evolving complex demands of the economy, many forward-thinking liberal arts institutions are actively and thoughtfully reimagining their curricula. They aim to address future workforce needs that combine the strengths of the liberal arts with practical, career-oriented components. For example, some universities, such as Carnegie Mellon University—Dietrich College, Wesleyan University, and Thapar School of Liberal Arts & Sciences, are now offering programmes that combine the rigorous study of philosophy with the analytical demands of data science, or the rich insights of sociology and literature with the practicalities of entrepreneurship. Such programmes are designed to produce graduates who are not only intellectually agile and deeply reflective but also ready to contribute on day one.

2.5 Structuring for Impact: An Exemplar Model

To operationalise these goals and counter the scepticism regarding workforce readiness, the Thapar School of Liberal Arts & Sciences has architected its programme around four distinct components. This structure is designed to move students systematically from broad interdisciplinary thinking to specialised application and research:

1. Interdisciplinary Foundation (Year 1): The first year is dedicated to breaking down academic silos. By immersing students in a broad range of disciplines, this component cultivates the intellectual agility required to forge connections across disparate fields.
2. Sustainability & Social Entrepreneurship Core (Semesters 2, 3, and 4): Running parallel to foundational coursework, this three-semester module ensures that students apply their learning to real-world problems focused on sustainability and social challenges. It embeds the necessity of ethical reasoning and societal well-being directly into the curriculum, forcing students to navigate real-world constraints.
3. Major and Minor Courses (Years 2 and 3): Once the interdisciplinary mindset is established, students deepen their technical and theoretical proficiency in specific and allied fields through their choice of major and minor studies. This stage mirrors the professional need for excellence seeking, ensuring graduates possess the deep subject-matter expertise required to complement their broad critical thinking skills.
4. Research (Year 4): The final year culminates in rigorous research. This component transitions students from consumers of knowledge to producers of evidence-based solutions, honing their capacity to analyse data and generate effective solutions—a critical asset in an age of information overload.

As technological advancement reshapes professional environments, educational programmes cannot remain static in their delivery methods or curricular focus. Students need preparation that enables them to complement, rather than compete with, technology-enhanced workplaces. The adaptability that forms the foundation of liberal arts education must extend beyond student learning to encompass institutional practice itself. Faculty and administrators must collaboratively embrace innovative teaching methods, integrate relevant technologies into coursework, and design experiences that mirror the collaborative, interdisciplinary nature of modern professional settings.

This institutional evolution requires more than surface-level changes. It demands a fundamental rethinking of how liberal arts skills are developed, practised, and applied. A few recommendations could be:

- curriculum: Integrate human–AI collaboration, data and algorithmic literacy, and applied ethics across courses. Treat prompting as brief writing and rhetoric; teach verification, attribution, and risk translation.

- pedagogy: Replace one-way delivery with studios, seminars, and labs that use current tools. Assess problem framing, audience fit, and judgment under uncertainty—not only final answers.
- practice: Expand work-integrated learning (client projects, internships, capstones), so students meet real constraints and standards of quality in contemporary workplaces.
- faculty development: Support instructors to adopt AI-enabled workflows and redesign assessments for integrity and learning.
- governance and incentives: Publish clear guidance on responsible AI use; reward co-teaching with technical units; align hiring and promotion with innovation in teaching and student outcomes.

These steps make liberal arts graduates better partners for technical teams and ensure that our quartet—communication, compassion, excellence-seeking, and lifelong learning—is practised in contexts that match the demands of the modern world.

2.6 Challenges and Opportunities

Despite the compelling and increasingly robust arguments for the relevance of a liberal arts education in the evolving future of work, several significant challenges persist. These are primarily centred around perceptions and prevailing policy frameworks. However, these very challenges also present opportunities for critical re-evaluation, innovation, targeted investment, and development of new pedagogical methods.

One of the most difficult challenges confronting liberal arts education today is the persistent perception gap. A significant segment of students, their parents, and even a notable portion of employers and governments continue to view liberal arts degrees as inherently less practical or career-oriented when compared to more vocational or STEM-focused degrees. This misconception is frequently aggravated by media narratives that tend to emphasise immediate job placement rates over the more nuanced and critical aspects of long-term career trajectory, adaptability, and leadership potential reflected in multiple studies and reports (Carnevale et al., 2020).

To effectively bridge this critical gap, liberal arts institutions must adopt a proactive and highly articulate approach to communicating the unique and multifaceted value proposition of their educational model and the advantages of interdisciplinary education. This involves not only highlighting the readily transferable skills—such as critical thinking, communication, genuine creativity, and sophisticated problem-solving—that are demonstrably highly sought after by employers across virtually all sectors, but also providing concrete evidence of their efficacy. Showcasing the diverse and impactful career paths of successful alumni who have adeptly leveraged their liberal arts backgrounds to thrive in a wide array of innovative and often unexpected careers can serve as exceptionally powerful testimonials.

Furthermore, deeper and more meaningful connections with industry partners, through well-structured internships, targeted career fairs, and active advisory boards, can significantly enhance employers' understanding and appreciation of the distinctive capabilities that liberal arts graduates bring to the workforce. A clear and consistent articulation of learning outcomes in terms of marketable competencies is also an absolutely crucial step in this strategic communication effort.

2.7 Conclusion

In today's world, characterised by unprecedented technological advancement and shifts in the global economy, the traditional, narrowly vocational model of education is proving insufficient to prepare individuals for the complexities and challenges posed by the future of work. This chapter has argued that a liberal arts education, empowered with technological advancements, offers a uniquely powerful framework for cultivating the essential skills and intellectual dispositions to truly thrive in this evolving landscape. By actively cultivating interdisciplinary reasoning skills, critical thinking capabilities, an empathetic mindset, and a capacity for adaptability and lifelong learning, liberal arts programmes equip individuals with rich competencies. These competencies are not only remarkably resistant to automation but are also increasingly in demand across virtually all sectors of the modern economy.

Reimagining the future of work through the nuanced lens of a liberal arts education necessitates a fundamental and collective shift in societal perception, coupled with a renewed and unwavering commitment to its foundational principles, and a redesign of programmes that address the challenges of the contemporary world. Educators, policymakers, employers, and students alike must collectively recognise that true preparedness for the twenty-first-century workforce extends far beyond narrow technical training. It demands a holistic and active intellectual agility, ethical reasoning, and understanding of the human condition.

By championing the liberal arts, we are not simply preserving a venerable academic tradition; we are, more importantly, actively investing in the creation of a future workforce that is not only highly skilled and remarkably adaptable but also deeply human, ethically grounded, and capable of navigating complexity, groundbreaking innovation, and contributing meaningfully to a more just, prosperous, and humane global society.

References

American Association of Colleges and Universities. (2023). *The career-ready graduates-ate: What employers say about the difference college makes*. Retrieved September 01, 2025, from https://dgmg81phhvh63.cloudfront.net/content/user-photos/Research/PDFs/AACU-2023-Employer-Report.pdf (National survey of 1,010 employers).

Autor, D. H., Levy, F., & Murnane, R. J. (2003). The skill content of recent technological change: An empirical exploration. *The Quarterly Journal of Economics, 118*(4), 1279–1333. https://doi.org/10.1162/003355303322552801

Carnevale, A. P., Cheah, B., & Van Der Werf, M. (2020). *Roi of liberal arts colleges: Value adds up over time*. Retrieved September 01, 2025, from https://cew.georgetown.edu/wp-content/uploads/Liberal-Arts-ROI.pdf

Choi-Lundberg, D., Douglas, T., Bird, M. L., Coleman, B., Greenwood, M., Martin, R., et al. (2024). Employability learning and teaching research: A twenty-year structured narrative review. *Journal of University Teaching and Learning Practice, 21*(5), 1–38.

Deming, D. J. (2017). The growing importance of social skills in the labor market. *The Quarterly Journal of Economics, 132*(4), 1593–1640. https://doi.org/10.1093/qje/qjx022

Edmondson, J., Formica, P., & Mitra, J. (2020). Special issue: Empathy, sensibility and graduate employment—Can the humanities help? *Industry and Higher Education, 34*(4), 217–220. https://doi.org/10.1177/0950422220928821

Han, S., LaViolette, J., Borkenhagen, C., McAllister, W., & Bearman, P. S. (2023). Interdisciplinary college curriculum and its labor market implications. *Proceedings of the National Academy of Sciences of the United States of America, 120*(43), 1–7. https://doi.org/10.1073/pnas.2221915120

Jackson, D., & Cook, E. J. (2025). Work-integrated learning in the humanities, arts and social sciences: Where to from here? *Studies in Higher Education, 50*(9), 2048–2067. https://doi.org/10.1080/03075079.2024.2409879

Lassébie, J., & Quintini, G. (2022). *What skills and abilities can automation technologies replicate and what does it mean for workers? New evidence*. OECD Social, Employment, and Migration Working Papers, (282), 0_1-66.

Lewis, P. (2018). Globalizing the liberal arts: Twenty-first-century education. In N. Gleason (Ed.), *Higher education in the era of the fourth industrial revolution* (pp. 15–38). Springer.

OECD. (2024). *Social and emotional skills for better lives: Findings from the OECD survey on social and emotional skills 2023*. OECD.

Sunstein, C. R. (1993). In defense of liberal education. *Journal of Legal Education, 43*, 22–26.

World Economic Forum. (2023). *The future of jobs report 2023*. Retrieved September 01, 2025, from https://www3.weforum.org/docs/WEFFuture of Jobs2023.pdf

World Economic Forum. (n.d.). *The future of jobs report*. World Economic Forum. https://www.weforum.org/reports/the-future-of-jobs-report-2023/

Zakaria, F. (2015). *In defense of a liberal education*. W. W. Norton & Company.

Chapter 3
Engineering Education Meets the Liberal Arts: Organising an Interdisciplinary Education for New Engineers

Pascal Wilhelm, Ardion Beldad, Jasper Homminga, Gustavo Adolfo García Chapeton, and Marcus Vinicius Pereira Pessoa

3.1 Introduction

Challenges facing the dynamic modern world are too complex to be addressed from a singular disciplinary perspective. This realisation has become increasingly apparent in engineering education (Van den Beemt et al., 2020). For this reason, engineering education has placed more emphasis on the development of team-based integrated problem-solving competencies (Passow & Passow, 2017). The disciplinary perspectives that are integrated should only not be limited to natural science and engineering domains, but must also include the social sciences and humanities. Apul and Philpott (2011) stated that interdisciplinary engineers should be able to handle and integrate environmental, social, and economic objectives in their endeavours because complex issues, such as ecological sustainability, require awareness of social, political, economic, and environmental constraints in problem-solving.

To mitigate the negative ramifications of complex technological and societal challenges and to help communities and societies become resilient, engineering

P. Wilhelm (✉)
Academy of Applied Psychology and Human Resource Management, Saxion University of Applied Sciences, Deventer, The Netherlands
e-mail: p.wilhelm@saxion.nl

A. Beldad · M. V. Pereira Pessoa
University College Twente, University of Twente, Enschede, The Netherlands

J. Homminga
University College Twente, University of Twente, Enschede, The Netherlands

Faculty of Science and Technology, University of Twente, Enschede, The Netherlands

G. A. García Chapeton
Academy of Life Sciences, Engineering, and Design, Saxion University of Applied Sciences, Enschede, The Netherlands

A. Beldad et al. (eds.), *Designing an Interdisciplinary Engineering Education for Complex Real-World Challenges*, Emergent Directions in Engineering Education 1, https://doi.org/10.1007/978-3-032-20223-9_3

education must train a new breed of engineers who can oversee the complexity of socio-technical systems and engage in collaborative problem-solving across disciplines. How could an academic bachelor's programme in engineering take shape that can meet these needs? That is the leading question of this chapter, in which the three-year bachelor's programme—Technology, Liberal Arts, and Sciences or ATLAS—at the University College Twente, University of Twente, the Netherlands, is described as a leading example of such a programme.

Started in 2013, this programme has been successful in implementing and integrating educational measures to meet these needs. The programme emphasises the integration of disciplinary knowledge (natural science, social science, and mathematics), academic and professional skills, and academic attitude. Knowledge integration and engagement in the processes of creation to tackle complex socio-technical challenges in an interdisciplinary setting require students to have an academic attitude and approach, and be competent in communication and collaboration.

From the outset, ATLAS was established for promising students who are not content with simple solutions to complex problems. The admission process of this selective programme screened for students who were not only prepared to enter an engineering programme, but also motivated by grand societal issues such as the scarcity of food, water, energy, and materials; the effects of population growth and urbanisation; and the implications of human behaviour for the environment. These students must be committed to finding solutions to, or at least reducing the impact of, these problems. The ATLAS programme is adjusted to these wider interests to ensure that students are sufficiently motivated and challenged.

To cater to these wide interests and foster interdisciplinarity, the programme adopted the Liberal Arts & Science framework. The 'LAS' in ATLAS embraces the social and behavioural sciences. The pure and engineering sciences are treated as an important subset of 'Technology' (hence the acronym ATLAS, in which the A is meaningless). ATLAS is the first Dutch university college to offer an interdisciplinary engineering programme with a distinct liberal arts character.

Liberal arts and science education has two important characteristics: exposure to a wide range of academic subjects and the use of a personal, student-centred approach (Becker, 2014). These characteristics were deemed suitable for the goals of fostering interdisciplinarity and nurturing talented students to become well-rounded system engineers.

Another important aspect to reach this educational goal was the adoption of problem-based learning and challenge-based learning. Student teams work on open and ill-defined real-world challenges and problems for a longer period (Van den Beemt et al., 2020). This fosters students' competence in dealing with complex issues in an interdisciplinary and collaborative setting.

To create a vision of the type of engineer the programme aimed to educate, five core values were introduced to steer the design of the programme. These core values distinguish the ATLAS programme from more traditional engineering programmes. The values express themselves in various ways. ATLAS students have a strong base in natural science, mathematics, and social science (interdisciplinarity). In semester projects, they work on challenges with a technology component (trailblazing), they live and work together (community), their programme is built on the principles of self-directed learning, and they design their curriculum to become the new engineer they want to be (self-directed learning and personal development).

The core values are also expressed in the final learning outcomes (FLOs) of the programme and further shaped by the academic criteria for bachelor's and master's curricula in engineering (Meijers et al., 2003) and the Domain Specific Frame of Reference (DSFR) that was developed for liberal arts and sciences programmes in the Netherlands. The DSFR framework aligns with European and international liberal arts and sciences networks, ensuring the connection with the expectations and requirements of the international educational field.

3.2 Final Learning Outcomes as Starting Points for Educational Programme Design

Preparing interdisciplinary engineering students for the complex challenges they will confront in the real world requires a curriculum that emphasises the holistic development of knowledge, skills, and attitudes. Describing the theories and models providing the foundations for interdisciplinary engineering education, Beldad and Miedema (2026) argue that interdisciplinary engineers must have the competencies indispensable for interdisciplinary collaboration, namely (a) a level of fluency in discipline-specific knowledge, (b) the ability to integrate a multiplicity of disciplinary knowledge for problem solving, (c) capacity to generate ideas and solutions, (d) collaboration and communication skills for multistakeholder cooperation, and (e) metacognitive skills for academic and personal growth (e.g., critical thinking, self-reflection). Based on this list of required skills that any interdisciplinary engineer must possess, the authors introduced a framework (see Fig. 3.1) for designing an interdisciplinary engineering education, which has served as the foundation for formulating ATLAS' final learning outcomes.

The framework emphasises strong integration among the four learning lines (disciplinary lenses for understanding, processes of creation, communication and collaboration, academic approach and attitude) of the ATLAS programme, with socio-technical challenges in the semester projects as an element steering the choice of contents and educational activities within a semester. Socio-technical challenges, such as the energy transition and improving well-being, stem from the complex interaction between technical and social factors (Walker et al., 2008), and they, understandably, require solutions informed by knowledge from the natural/technical and social science domains (Savaget et al., 2019).

The first point of the framework is that the contents selected in the disciplinary lenses for understanding (natural science, social science, and mathematics) are predicated on the nature of the socio-technical challenges. The second point's emphasis on pursuing the processes of creation is a response to calls for engineering professionals to develop sound and responsible solutions through research and

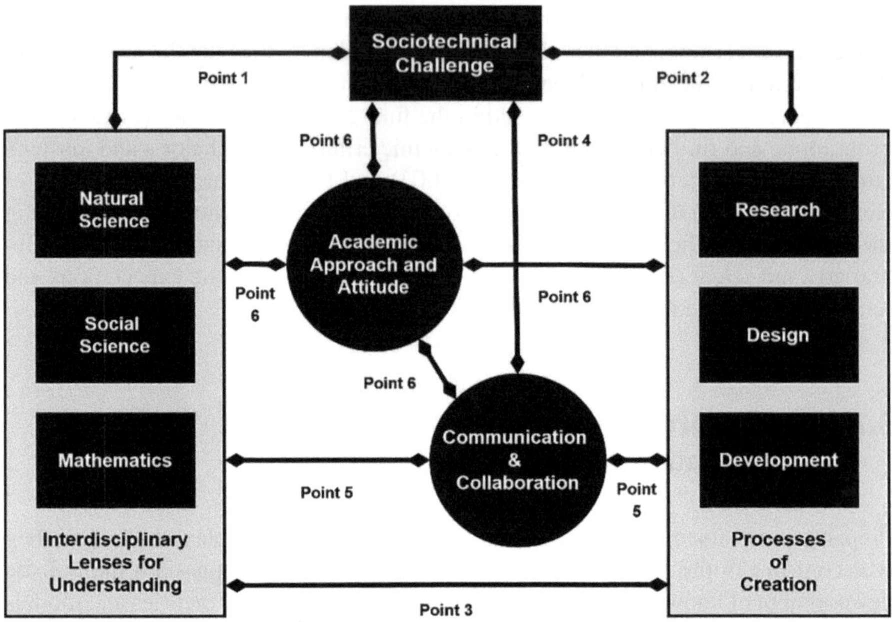

Fig. 3.1 Framework for designing the ATLAS educational programme as proposed by Beldad and Miedema (2026)

design. The third point is founded on the premise that educational activities in discipline-specific courses facilitate students' acquisition and development of valuable research and design skills.

The experience of interacting and collaborating with different stakeholders when tackling sociotechnical challenges facilitates the development of ATLAS students' communication and collaboration skills, as highlighted in point 4. Point 5 highlights the importance of designing learning activities across various courses to hone students' communication and collaboration skills, while their research and design projects provide more opportunities to develop these skills.

The centrality of the academic approach and attitude in the framework, as accentuated in point 6, signifies that while metacognitive (e.g., critical and analytical thinking) and ethical reflection skills are imperative for students to perform well in courses and project works, students' learning activities throughout a semester are fitting opportunities for cultivating their academic mindset.

The four learning lines presented in the framework partly concur with what Schijf et al. (2023) claim as necessary ingredients for developing students' interdisciplinary mindset, namely (a) knowledge of different disciplinary paradigms and interdisciplinary ways of working, (b) having a critical and reflective attitude, and (c) proficiency in communicating and collaborating with different actors. In the framework used by ATLAS, point A pertains to the 'disciplinary lenses for understanding', while point B is the 'academic approach and attitude' in the framework. The inclusion of the 'processes of creation' conforms to the recommendations of the Conceive, Design, Implement, Operate (CDIO) Framework of educating engineering students to have in-depth knowledge of technical topics, to lead in creating and operating novel products and processes, and to understand the societal implications of research and technological developments (Crawley et al., 2014).

The emphasis on the four learning lines reflects the ATLAS programme's vision of educating interdisciplinary engineers who are not only proficient in the languages of the different sciences, and competent in integrating various disciplinary knowledge to address complex socio-technical challenges, but also adept in designing solutions and producing new knowledge, and competent in interacting and collaborating with different social actors. For each learning line, final learning outcomes (FLOs) are specified to function as starting points for the formulation of learning goals at the semester and course levels. This approach adheres to Thijs and Van den Akker's (2009) curricular spider web components, which involve specifying the reasoning and the goals of the educational programme. The aims indicated inform the contents of education and their associated learning activities. Table 3.1 shows the correspondence between the learning lines and the final learning outcomes of the ATLAS programme.

3.3 Facilitating Students' Attainment of ATLAS's Final Learning Outcomes

The three-year ATLAS programme is organised in six semesters, with the first four semesters dedicated to a specific socio-technical challenge. The fifth semester provides students with the opportunity to pursue their educational activities at another university, either abroad or in the Netherlands. The sixth semester is designed for students to complete a Capstone project—either research- or design-oriented—while also taking elective courses relevant to their intended academic profile.

After the first semester, students are already encouraged to focus on an academic profile, similar to the idea of students deciding on a specialisation in either natural science, social science, or engineering within a liberal arts programme. This academic direction prompts students to be deliberate in designing their educational curricula, an approach afforded by the ATLAS programme's open character, which, as Kovačević et al. (2024) claim, is one of the defining features of a liberal arts education.

Except for the first semester, in which all learning activities for reaching the semester learning goals are prescribed, semesters 2 to 6 provide students with enough space to identify and select educational activities to realise those goals. Semester goals are deliberately formulated broadly to allow students to personalise them by specifying the learning activities they will pursue to achieve each goal and the types of evidence they will collect to demonstrate successful goal attainment. This requirement of learning goal personalisation is emphasised throughout the educational programme to nurture students' sense of ownership and control of the

Table 3.1 Correspondence between the ATLAS' Learning Lines and the programme's final learning outcomes (FLOs)

Learning line	Statement of ATLAS Final Learning Outcomes Successful completion of the three-year ATLAS programme signifies that a student:
Disciplinary lenses for understanding	(a) Understands the knowledge base and the concepts, laws, principles, and models in mathematics, natural science, and social science;
	(b) Knows how theories and models emerge and evolve and how they can be used to understand phenomena in relevant fields;
Processes of creation	(c) Is able to create solutions for socio-technical challenges based on systematic approaches involving research, design, and development;
	(d) Is able to apply, justify, and integrate theories and models from the natural sciences, social sciences, and mathematics, and use them adequately in creating solutions for socio-technical challenges;
Communication and collaboration	(e) Is able to professionally, responsibly, and respectfully collaborate with peers and other professionals with different social and academic backgrounds and in different settings to attain specific goals and objectives;
	(f) Is able to appropriately communicate information, insights, standpoints, and new knowledge in various ways to different types of audiences;
	(g) Is able to engage in both academic and non-academic discussions on information, insights, standpoints, and new knowledge with various groups and individuals in a socially and culturally sensitive manner;
Academic approach and attitude	(h) Has the self-awareness to make well-informed decisions concerning their academic and personal growth by identifying appropriate strategies to continuously develop their competencies and skills, and to evaluate and reflect on their learning processes and gains;
	(i) Acts in a socially responsible manner and can reflect on the societal and ethical implications of research, design, and development; and
	(j) Is inquisitive and can ask relevant and critical questions, as well as find answers to those questions in an academically sound manner

learning process, as this has been shown to bolster students' intrinsic motivation to learn (Mikalayeva, 2016; Valerio, 2012).

Through a broadly formulated learning goal, such as 'to demonstrate academic competencies in fields or domains that constitute your profile as a new engineer at an advanced level', students can specify educational activities (e.g., selecting elective courses) that would allow them to acquire disciplinary competencies relevant to their chosen academic profile or direction. This kind of learning goal personalisation requires students to identify their learning needs, specify ways to satisfy those needs, and decide on how to determine successful learning goal attainment.

The benefits of offering students autonomy and choice in designing their educational curriculum, however, are also countered by potential risks, such as having activities incongruent with the programme's learning lines (hence, not enabling the student to fulfil the programme's final learning outcomes) and students selecting courses well below the expected level. To mitigate these possible risks, ATLAS implements three mechanisms.

First, in semesters with mandatory educational activities (e.g., obligatory courses, semester project), assignments are judiciously designed to ensure that students can attain the learning goals associated with the four learning lines. For instance, a research project in a social science course must not only enable students to experience the processes of creation (e.g., knowledge production through research), but must also enable them to engage with the scientific literature to understand a phenomenon prompting the research and to pursue collaborative research activities, communicate their research findings, and develop metacognitive (e.g., critical thinking) and ethical reflection skills.

Second, specifying requirements for identifying educational activities is critical. Students in the higher semesters (semesters 4 to 6) are expected to take elective courses at the second- or third-year bachelor's level, or even at the master's level. Such a requirement helps guarantee that the curricular choices made by students will enable them to acquire the competencies needed to strengthen their academic profile.

Third, the autonomy students have in designing their curricula is complemented by support and guidance from academic staff members in the three phases of the programme's self-directed learning (SDL) cycle. At the start of a semester, students are required to write a personal development plan (PDP) that describes their learning needs for each learning goal of a semester, the specific educational activities (e.g., courses) they will pursue to realise those needs, and the evidence that will be collected to show successful attainment of the learning goals. This primarily subscribes to the principle of learning goal personalisation previously mentioned.

Halfway through a semester, a midterm evaluation (MTE) is scheduled with a student and a pair of academic staff to check the student's progress in meeting the learning goals according to the educational activities specified in the PDP. This MTE is also a fitting opportunity for students to revisit their original plans to see if any adjustment to the plan is warranted, and for the academic staff to guide students in identifying alternative courses of action when students' approaches for meeting the learning goals are not delivering the expected results. The MTE concludes with an indication from both the student and academic staff of whether or not the student is on track to realise the semester's learning goals.

When deemed necessary, the student revises the PDP, which academic staff within a semester must approve before the self-evaluation and reflection (SER). Approval of the PDP signifies that the educational programme officially recognises the academic quality of a student's curriculum, resulting in the registration of the results (e.g., grades) of the student's educational activities (e.g., courses) in an official transcript of records.

Scheduled towards the end of a semester, and when all the learning activities of students have been completed, the SER is an open discussion between a student and a pair of academic staff. In the first part of this discussion, students evaluate the results of all the educational activities specified in the PDP to determine the extent to which the activities and their results enable the attainment of the semester's learning goals (the self-evaluation part of the SER). In the second part of the discussion, students reflect on the implications of their educational activities and results for their academic and personal growth, as well as on their plans for the succeeding semester (the reflection component of the SER).

Mentoring, guidance, and feedback provision throughout the self-directed learning cycle are crucial in an educational programme that emphasises personal development and self-directed learning. For one, the freedom provided by SDL must be regulated by curricular and professional requirements (Iwasiw, 1987). Secondly, significant variations in students' self-directed learning readiness (Fisher et al., 2001) and their propensity to increasingly take responsibility for the learning process (Hewitt-Taylor, 2001) have been reported. Hence, the programme espouses a balance between giving students autonomy over their learning and providing them with academic guidance to help them make informed choices when designing their curriculum. Additionally, as Chan et al. (2014) claim, the provision of feedback by academic staff throughout the SDL cycle is instrumental in aiding students to monitor their progress and growth.

3.4 Semesters as Building Blocks of ATLAS Students' Interdisciplinary Education

An academic year in the ATLAS programme consists of two main components: two semesters (worth 27 European Credits or ECs each, with 1 EC equivalent to 28 study hours) and one Personal Pursuit (worth 6 ECs). The semesters run for 20 weeks and are considered the main building blocks of the curriculum. Chapter 8 by Dijksterhuis describes the organisation and requirements of the Personal Pursuit.

A semester structure (as opposed to the university's quartile structure) is used in the ATLAS programme, as the first four semesters contain a socio-technical project (worth 9 ECs). A semester duration of approximately 20 weeks ensures that students can work on their projects over a relatively long period, allowing for sufficient depth and breadth.

In the projects, students apply their theoretical knowledge, which can deepen their understanding of materials from mandatory courses. At the same time, they engage in the processes of creation (through research and design), particularly since the first three projects have a design orientation, while the fourth semester immerses students in the complete research cycle. The projects build on each other and gradually increase in complexity. The first-semester project focuses on solving problems at the individual level, the second-semester project addresses a problem at the community level, and the third-semester project tackles a wicked problem at either the national or international level.

Alongside the semester project are courses that are either mandatory or elective. The first three semesters contain mandatory courses for the 'lenses for understanding' (natural science, social science, mathematics). The mandatory courses are designed to provide students with foundational disciplinary knowledge. These mandatory courses also serve as starting points when students are building their academic profile. Integration of knowledge from these mandatory courses into the semester projects is expected to help students gradually develop interdisciplinary competencies.

The mandatory courses decrease in size over the semesters, from two-thirds of the first semester to one-third of the second and third semesters. This reduction in mandatory courses is complemented by an increase in elective courses, through which students can develop their own academic profile. Students can take these elective courses in any of the other educational programmes at the University of Twente or even abroad, when students reach the educational exchange semester (semester 5).

If students cannot find the courses they need, they can develop their own course in cooperation with any member of the programme's academic staff. This newly developed course constitutes a capita selecta, which displays students' ability to direct their learning, especially as starting a capita selecta requires students to identify their learning goals and needs, specify the knowledge they intend to acquire, indicate the product or assignment that will be completed, and propose a method for assessing the deliverable.

The first semester is fully predetermined, which is a deliberate curricular choice to give students enough time to get used to university life and to new ways of learning before they start developing their academic profile. In the fifth semester, students have opportunities to experience other types of education in new learning environments by joining an academic exchange. At this point, students have the option to follow elective courses abroad or in other Dutch universities. A few will eventually decide to do educational modules in other programmes within the University of Twente. This is the semester when students are already in the mature phase of building their academic profiles. Some students might even opt to spend the fifth semester doing an internship to acquire more professional skills, gain insights into the actual world of work, and have opportunities for knowledge application in professional settings.

During the sixth semester, students are expected to finalise their academic profile by completing a Capstone project, either with a design or research orientation. Through this project, students can demonstrate the design and research skills they have acquired from the previous semesters. The semester also provides some flexibility for students to take elective courses relevant to their academic profile. Figure 3.2 illustrates the structure of the six semesters of the ATLAS programme, with the various elements of each semester represented by different shades of black and grey, accompanied by textual labels.

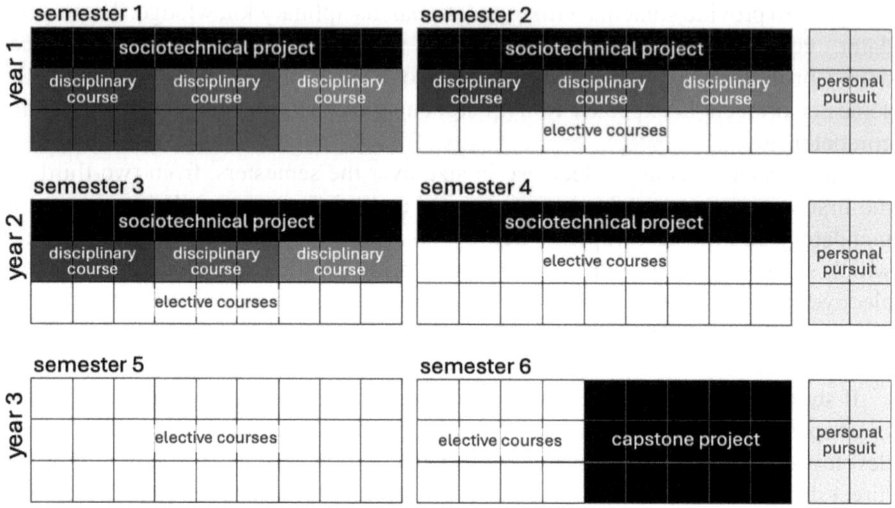

Fig. 3.2 Structure of the ATLAS curriculum (the central socio-technical projects, the disciplinary courses, and the Personal Pursuit are deliberately distributed over the three years of the programme)

3.5 Designing a Semester Project: Exposing Students to Real-World Problems

Semester projects in the ATLAS programme are organised according to the principles of problem-based learning (PBL) and challenge-based learning (CBL). As an educational approach, PBL primarily involves students constructing an end product (a 'concrete artefact') that displays both knowledge and skills acquired from courses and their ability to apply knowledge to solve a real-world problem (Kokotsaki et al., 2016). CBL is a pedagogical approach that engages students with open-ended, real-world challenges, requiring the adoption of multidisciplinary perspectives, self-directed collaborative work, and critical and creative thinking (Leijon et al., 2022).

While the two pedagogical approaches differ in their emphases, both promote active, self-directed, and collaborative learning. In the context of ATLAS semester projects, a key difference lies in how specific (as in PBL) or open-ended (as in CBL) the project scope is. The projects are intentionally designed to transition gradually from designing a specific product or system in the early semesters to more open-ended challenges in later semesters. This progression is illustrated in Fig. 3.3. These pedagogical approaches align closely with the core values of the ATLAS programme and provide a pedagogical foundation to design a semester around a project.

Two examples of semester projects, representing different levels of problem complexity and scope, are described below.

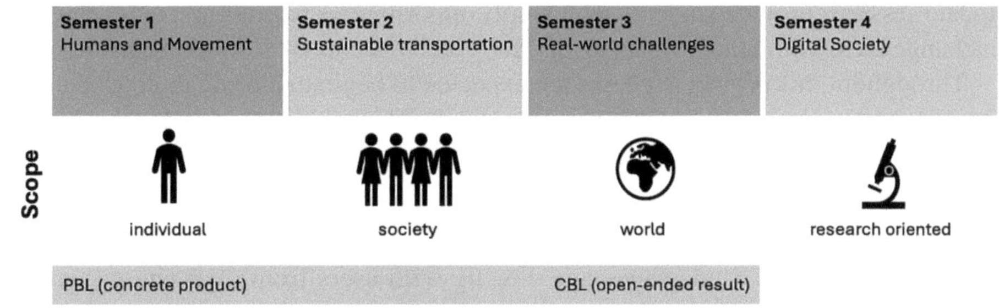

Fig. 3.3 Evolution of the projects across the first four semesters of the programme

3.5.1 Semester 1 Project 'Introduction to Sociotechnical Design'

The first semester project is organised according to the principles of design thinking and PBL. As an educational approach, design thinking is a human-centred, iterative process for creative problem-solving involving five phases, namely empathising, defining, ideating, prototyping, and testing (Brown, 2008). This approach aligns closely with PBL's emphasis on students constructing concrete artefacts that display both knowledge and skills acquired from courses and their ability to apply knowledge to solve real-world problems (Kokotsaki et al., 2016).

As the cornerstone of the semester on Humans and Movement, the project is designed to allow students to apply disciplinary knowledge from mandatory courses when developing specific solutions. The knowledge and skills acquired from those courses and other educational activities enable students to (a) understand the fundamentals of engineering as the act of working artfully to bring something about, (b) experience the transition from secondary education to university-level study, (c) develop multidisciplinary competencies for socio-technical design, and (d) cultivate collaborative teamwork skills.

Since ATLAS students have a higher level of autonomy in their academic choices than students from more traditional programmes, this first project is intentionally designed to provide a structured yet exploratory introduction to engineering practice. The theme of Humans and Movement, while broad in its definition, highlights aspects of quality of life that students can readily connect with through observation and empathy. In this context, students are tasked with designing a physical product for a specific target group, either elderly individuals or people with special needs. These products are expected to assist the chosen group in performing their daily tasks.

In the project, students work in teams of four to six members, learning to navigate the transition from secondary school to self-directed university study. The teams are partially self-organised, with tutors providing structured guidance in project management and socio-technical design processes. Tutors with expertise in engineering and social sciences serve as the primary points of contact. Regular update sessions provide students with ample opportunities to present their progress, exchange ideas with other project teams, and receive feedback from peers and tutors.

Throughout this project, students are expected to begin exploring their academic interests and apply essential teamwork competencies, including task planning, communication, role distribution, and managing design iterations and user feedback. The design thinking framework structures the team's work into clearly defined phases, each with specified activities and expected outcomes:

- Empathise phase: Students engage directly with users from their chosen target group, conducting observations and interviews to understand target users' needs, challenges, and daily routines.
- Define phase: The team synthesises insights from the empathise phase to articulate a clear problem statement that will guide the design process.
- Ideate phase: Students generate multiple creative concepts for addressing the identified problem, exploring diverse solution pathways.
- Prototype phase: The team develops tangible prototypes of their most promising design concepts, ranging from low-fidelity mock-ups to functional models.
- Test phase: Students evaluate their prototypes with users, gathering feedback to refine their designs and assess the viability of proposed solutions.

Although tutors provide structured guidance throughout the project, students are encouraged to take ownership of their design process. Domain courses offer foundational knowledge in relevant engineering and social science concepts, while design workshops introduce methods for user research, prototyping, and testing. Expert lectures provide insights into professional practice, while communication workshops develop students' ability to articulate design decisions. Students benefit from access to active project spaces in the ATLAS building and the University of Twente's Design Lab, which foster hands-on experimentation and collaborative work.

A concrete example of the Semester 1 Project was the collaboration with an elderly care organisation in Enschede, conducted between September 2023 and January 2024. Students were paired with elderly volunteers who agreed to participate in the project, working under the supervision of care professionals within a strict ethical framework that protected participant privacy and dignity.

This direct engagement with real stakeholders proved transformative for student learning. Teams developed solutions addressing authentic daily challenges, such as:

- an assistive device for putting on and removing coats independently,
- an ergonomic tool for opening food containers with reduced grip strength,
- an extendable reaching mechanism for accessing high shelves or low sink bottoms, and
- a stability support system for safe movement within living spaces.

The experience of interacting with users whose quality of life would be directly impacted by their work significantly boosted student engagement and motivation. Students reported that knowing their designs could make a meaningful difference in someone's daily life fostered deeper collaboration within teams and propelled them to develop more thoughtful and creative solution concepts. The resulting prototypes demonstrated not only technical competence but also genuine empathy for user needs and an emerging understanding of the socio-technical nature of design challenges.

3.5.2 Semester 3 Project 'Real-World Challenges in Complex Real-World Spatio-Temporal Systems'

Similar to the semester 1 project, the project in the third semester also facilitates student application of knowledge and skills from previous semesters and the current semester when (a) studying and analysing a real-world challenge, (b) identifying the relevant stakeholders and critical issues, (d) formulating and implementing mitigation strategies, and (d) reflecting on how their work benefits society.

Figure 3.4 shows the rationale behind the design of the Semester 3 project. In this project, students assume the role of professionals working within an international, multidisciplinary team of 8 to 15 members, tasked with addressing a real-world challenge (e.g., wildfires in the Mediterranean or illegal drug trafficking in Europe). The teams are self-organised and self-directed, with teachers within the semester

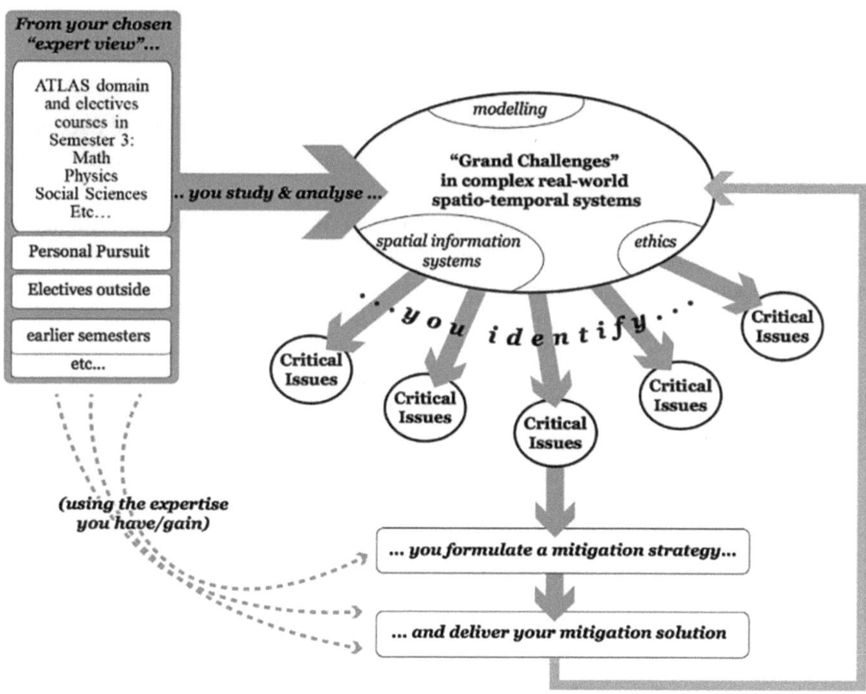

Fig. 3.4 Design rationale of the Semester 3 project

acting solely as advisors. One student is appointed to lead the project team in meeting its targets.

The team is expected to thoroughly understand the problem, identify the most pressing issues to be addressed, and propose mitigation strategies that meaningfully integrate knowledge from both the natural sciences and the social sciences. Throughout this project, students are expected to further develop their academic profiles and develop and apply essential teamwork competencies, including project planning, communication, task delegation, resource management, and managing conflicts and uncertainties.

Working on a real-world challenge is a complex task that can easily become overwhelming. To provide a good starting point for the student, the project follows a well-defined framework, as illustrated in Fig. 3.5. The project is divided into four stages, each with specified timelines and expected outcomes.

- Preparation phase: the team (referred to in the figure as the *large team*) is formed, and the challenge to be addressed is selected.
- Mapping & Modelling phase: the team collects and analyses information related to the challenge and prepares a report to communicate the results of the analysis.
- Mitigation phase: building on the critical issues identified during the Mapping & Modelling phase, the team splits into smaller sub-teams (of up to three members each) to propose and implement actions aimed at mitigating the negative effects of the selected critical issues.
- Integration phase: the large team synthesises the outcomes and reports on the expected new situation, such as the expected changes resulting from the proposed mitigations.

Although the teacher adopts a largely advisory role in the planning and execution of the team's work, the required deliverables serve as formal checkpoints to ensure that progress remains on track.

A concrete example of the Semester 3 project is the study of earthquakes in Turkey. During the Mapping & Modelling phase, the team identified eleven societal problems associated with earthquakes, seven critical issues, and 44 individual stakeholders, which were organised into five categories.

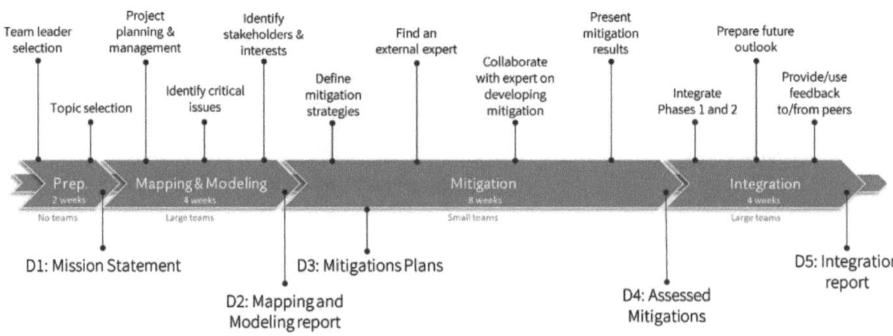

Fig. 3.5 Semester 3 project structure

The Mitigation phase resulted in four proposed mitigation strategies.

- Development of a protocol for conducting Geographic Information Systems (GIS)-based evaluations of critical infrastructure and road network connectivity, supporting the creation of response plans based on projected infrastructure damage.
- Recommendations for the use of earthquake-resilient materials in school buildings, aiming to reduce casualties caused by structural collapse.
- Design of a locomotion mechanism for search-and-rescue robots in disaster areas, leading to the creation of a working prototype.
- A study of the psychological effects of post-earthquake trauma to inform strategies for enhancing psychological resilience in affected populations.

Finally, the Integration phase resulted in a comprehensive analysis of how the identified stakeholders and potential mitigations could collectively improve earthquake preparedness and reduce the societal costs of these unavoidable natural events.

3.6 Conclusion

In this chapter, it is argued that future engineering education should foster interdisciplinarity and team-based integrated problem-solving competency. This is necessary to face modern-day challenges that are simply too complex to be addressed from a singular disciplinary perspective. The leading question is how such a programme could take shape. An example of such a programme (the ATLAS programme of the University College Twente, Univesity of Twente) is presented that integrates the educational approaches of liberal arts and sciences and self-directed learning and introduces core values (interdisciplinarity, community, trailblazing, personal development, and self-directed learning) to shape the profile of future engineers. These core values are translated into Final Learning Outcomes (FLOs) that adhere to current frameworks of bachelor's and master's curricula in engineering (Meijers et al., 2003) and international Liberal Arts and Sciences curriculum networks.

From the FLOs, learning lines extend across the curriculum to define the contents of the programme. These learning lines include disciplinary lenses for understanding, processes of creation (design and research), communication and collaboration, and academic approach and attitude (e.g., personal growth, social responsibility, critical inquisitiveness). A central component of the programme is the promotion of interdisciplinarity and integrated team-based problem-solving in semester projects. Two such projects are described in more detail, highlighting the context of complex, open-ended challenges and the team-based, integrated mitigation approach that students are learning. Another unique feature of the programme is the opportunity for students to build their own academic profiles as future engineers.

The ATLAS programme is by no means the only way an interdisciplinary engineering programme for future engineers could take shape. Even when one would adopt the same framework that served as a mould to shape the programme (e.g., liberal arts and sciences, project/challenge-based learning), a different programme may emerge. However, adopting relevant frameworks that fit future needs in engineering education and continuously using them as touchstones to steer the continuous development of a programme certainly helps in doing things right (the *how*) as opposed to doing the right thing (the *what*). The latter refers to what is clearly a need in engineering education—to tear down disciplinary silos and let interdisciplinarity be the leading principle in redesigning engineering education.

Acknowledgement The authors would like to thank Barend Köbben from the Faculty of Geo-Information Science and Earth Observation, University of Twente, for allowing us to use Fig. 3.4 of this chapter, which illustrates the rationale for the design of the semester 3 project of the ATLAS programme. Köbben was primarily responsible for the semester project on grand challenges when the ATLAS programme started.

References

Apul, D. S., & Philpott, S. M. (2011). Use of outdoor living spaces and Fink's taxonomy of significant learning in sustainability engineering education. *Journal of Professional Issues in Engineering Education and Practice, 137*(2), 69–77. https://doi.org/10.1061/(ASCE)EI.1943-5541.0000051

Becker, J. (2014). *What a liberal arts and sciences education is... and is not* (Final September 2014) [PDF]. Bard College. https://tools.bard.edu/wwwmedia/files/5438940/1/Liberal%20Arts%20and%20Sciences%202014%20Final%20September%2014.pdf.

Beldad, A. D., & Miedema, H. A. T. (2026). Introducing a framework for designing an interdisciplinary engineering curriculum: Educating new engineers for complex sociotechnical challenges. *European Journal of Engineering Education, 51*(1), 192–209. https://www.tandfonline.com/doi/full/10.1080/03043797.2025.2507243

Brown, T. (2008). Design thinking. *Harvard Business Review, 86*(6), 84–92.

Chan, P. E., Graham-Day, K. J., Ressa, V. A., Peters, M. T., & Konrad, M. (2014). Beyond involvement: Promoting student ownership of learning in classrooms. *Intervention in School and Clinic, 50*(2), 105–113. https://doi.org/10.1177/1053451214536039

Crawley, E. F., Malmqvist, J., Östlund, S., Brodeur, D. R., & Edström, K. (2014). *Rethinking engineering education*. Springer.

Fisher, M., King, J., & Tague, G. (2001). Development of a self-directed learning readiness scale for nursing education. *Nurse Education Today, 21*(7), 516–525. https://doi.org/10.1054/nedt.2001.0589

Hewitt-Taylor, J. (2001). Self-directed learning: Views of teachers and students. *Journal of Advanced Nursing, 36*(4), 496–504. https://doi.org/10.1046/j.1365-2648.2001.02001.x

Iwasiw, C. L. (1987). The role of the teacher in self-directed learning. *Nurse Education Today, 7*(5), 222–227. https://doi.org/10.1016/0260-6917(87)90005-0

Kokotsaki, D., Menzies, V., & Wiggins, A. (2016). Project-based learning: A review of the literature. *Improving Schools, 19*(3), 267–277. https://doi.org/10.1177/1365480216659733

Kovačević, M., Dekker, T. J., & van der Velden, R. (2024). Employability development in undergraduate programmes: How different is liberal arts education? *Teaching in Higher Education, 29*(8), 2184–2204. https://doi.org/10.1080/13562517.2023.2212602

Leijon, M., Gudmundsson, P., Staaf, P., & Christersson, C. (2022). Challenge-based learning in higher education—A systematic literature review. *Innovations in Education and Teaching International, 59*(5), 609–618. https://doi.org/10.1080/14703297.2021.1892503

Meijers, A. W. M., Van Overveld, C. W. A. M., & Perrenet, J. C. (2003). *Academische criteria voor bachelor en master curricula = Academic criteria for bachelor and master curricula.* Technische Universiteit Eindhoven.

Mikalayeva, L. (2016). Motivation, ownership, and the role of the instructor in active learning. *International Studies Perspectives, 17*(2), 214–229. https://doi.org/10.1093/isp/ekv001

Passow, H. J., & Passow, C. H. (2017). What competencies should undergraduate engineering programs emphasize? A systematic review. *Journal of Engineering Education, 106*(3), 475–526. https://doi.org/10.1002/jee.20171

Savaget, P., Geissdoerfer, M., Kharrazi, A., & Evans, S. (2019). The theoretical foundations of sociotechnical systems change for sustainability: A systematic literature review. *Journal of Cleaner Production, 206*, 878–892. https://doi.org/10.1016/j.jclepro.2018.09.208

Schijf, J. E., Van der Werf, G. P., & Jansen, E. P. (2023). Measuring interdisciplinary understanding in higher education. *European Journal of Higher Education, 13*(4), 429–447. https://doi.org/10.1080/21568235.2022.2058045

Thijs, A. & Van Den Akker, J. (2009). *Curriculum in development.* Netherlands Institute for Curriculum Development (SLO).

Valerio, K. (2012). Intrinsic motivation in the classroom. *Journal of Student Engagement: Education Matters, 2*(1), 30–35.

Van den Beemt, A., MacLeod, M., Van der Veen, J., Van de Ven, A., Van Baalen, S., Klaassen, R., & Boon, M. (2020). Interdisciplinary engineering education: A review of vision, teaching, and support. *Journal of Engineering Education, 109*(3), 508–555. https://doi.org/10.1002/jee.20347

Walker, G. H., Stanton, N. A., Salmon, P. M., & Jenkins, D. P. (2008). A review of sociotechnical systems theory: A classic concept for new command and control paradigms. *Theoretical Issues in Ergonomics Science, 9*(6), 479–499. https://doi.org/10.1080/14639220701635470

Chapter 4
Striking the Balance between Freedom and Confinement: Implementation of Self-Directed Learning in an Interdisciplinary Bachelor's Engineering Programme

Pascal Wilhelm

4.1 Introduction

In self-directed learning, learners are put in the driver's seat of their own learning process. Learners diagnose their own learning needs, formulate goals, identify resources, select and implement learning activities, and evaluate their own learning outcomes (Knowles, 1975, see also Morris, 2019). According to Hiemstra and Brockett (2012), self-directed learning can lead to an increased sense of empowerment in students to take responsibility for their own learning. Knowles (1975) mentions several reasons why self-directed learning is relevant.

First, by taking an active stance towards one's own learning, learning becomes more effective. Although the evidence on its effectiveness is mixed, active learning strategies are commonly preferred over more passive ones (see Prince, 2004). Secondly, self-directed learning aligns well with our natural psychological development, which is focused on becoming increasingly independent from one's caregivers. Third, learning approaches in higher education (e.g., inquiry learning, project-based learning, or challenge-based learning) increasingly call upon the learners' ability to self-direct their learning.

Although put forward in the 70s by Knowles, self-directed learning is still relevant today. What is more, and Knowles showed foresight in this respect, education is not finished after learners finish their formal education. They need to continuously update their knowledge and skills to be able to contribute to society until they leave the workforce. Lifelong learning is considered essential in a rapidly changing

P. Wilhelm (✉)
Academy of Applied Psychology and Human Resource Management, Saxion University of Applied Sciences, Deventer, The Netherlands
e-mail: p.wilhelm@saxion.nl

A. Beldad et al. (eds.), *Designing an Interdisciplinary Engineering Education for Complex Real-World Challenges*, Emergent Directions in Engineering Education 1, https://doi.org/10.1007/978-3-032-20223-9_4

world (Morris, 2019, 2024). As Knowles (1975, p. 17) puts it: "We are talking about a basic human competence; the ability to learn on one's own—that has suddenly become a prerequisite for living in this new world."

Taken together, self-directed learning is an approach that seems to foster students' motivation and learning outcomes and provides a basis for lifelong learning at the same time. Murray (2015) discussed the interrelationships between self-directed learning and lifelong learning. There is a common consensus that higher education can prepare learners for lifelong learning by implementing self-directed learning approaches (Morris, 2019, 2024).

4.2 The Challenge with Implementing Self-Directed Learning in Higher Education

Although more "open" approaches to learning are implemented in higher education that call upon skills in self-directed learning, study programmes still contain mandatory courses in which teachers teach and students must show they master what is taught by doing assignments, taking tests, and exams. The basic model for an educational programme is still that a set of programme outcomes needs to be attained by completing a set of mandatory and elective courses with formal exams. There is nothing inherently wrong with such a system (it seemed to have served society well), but it still stands at a considerable distance from a system that fully embodies self-directed learning. Students are still required, to a great extent, to learn what the study programme wants them to learn, in a way the study programme wants them to learn, and the study programme decides based on their tests and exams whether the students have attained the study programme's learning goals.

University education requires adherence to national and local rules and regulations, educational missions and visions, and quality standards for education and educational assessment. In addition, there is still a need for formal educational qualification systems that allow employers to learn what new employees are capable of. These requirements do not easily fit with the principles of self-directed learning, in which the learners decide what they want to learn, in what way they want to learn, and evaluate for themselves whether they managed to learn what they set out to learn. To move a bit higher on the continuum of self-direction, as Hiemstra and Brockett (2012) put it, how can self-directed learning be implemented to a fuller extent, given its potential and its importance for lifelong learning, but also given the current requirements and frameworks higher education programmes need to adhere to?

4.3 Taking the Challenge: The ATLAS Programme

This was the challenge the ATLAS programme (Technology, Liberal Arts and Sciences) of the University College Twente (part of the University of Twente in the Netherlands) was facing. This three-year bachelor's engineering programme was inspired by the idea that new technologies are crossing traditional disciplinary boundaries, which implies that engineering solutions require an integrated perspective, especially in the way technologies are introduced into and affect society (see e.g., Bammer, 2013).

In line with this idea, the ATLAS programme was inspired by what Goldberg et al. (2014) coined as "The New Engineer". For ATLAS, a visionary profile for this New Engineer was created that includes five core values (UCT, 2018). These include interdisciplinarity, trailblazing, community, self-directed learning, and personal development. These core values stipulate what sets ATLAS graduates apart from graduates from traditional engineering programmes. The core value of self-directed learning is defined as: "ATLAS graduates have full awareness of their strengths and limitations and are open to various opportunities and approaches for learning." Personal development is defined as: "ATLAS graduates are intrinsically motivated to develop themselves holistically and can critically reflect on their growth."These two core values are expressed in the adoption of the self-directed learning concept as the basis for curriculum design, the ability to steer one's own learning is also considered an important skill for New Engineers to develop, which is in line with Knowles (1975) and Morris (2019, 2024).

Being able to handle complex challenges in multidisciplinary contexts in a society in which technological innovation accelerates at a very high rate requires the necessity for lifelong learning (Boyer, 2014; Morris, 2019; Loeng, 2020). The implementation of self-directed learning in ATLAS aims at creating a positive attitude towards continuous personal and professional development. Although this attitude is not addressed directly, it is assumed that by experiencing a self-directed learning approach for three years or more, this attitude is created and nurtured and will transfer to learning behaviour in professional contexts. In another chapter in this book, a study is presented that explored to what extent having experienced a self-directed approach in higher education affects learning attitudes and behaviours post-education (Kampouroglou et al., 2026, this volume).

The central question this chapter addresses is: how to design a curriculum that is fully based on self-directed learning while still adhering to the frameworks for designing good quality higher education? The design of the ATLAS programme is used to illustrate how this challenge can be faced.

4.4 Implementing Self-Directed Learning in ATLAS: Garrison's Model of Self-Directed Learning

Garrison (1997) presents an integrative, comprehensive model of self-directed learning comprising self-management, self-monitoring, and motivation as main dimensions. He argues that these dimensions cannot be seen as separate entities, but that they are interrelated. The author takes a collaborative constructivist perspective on self-directed learners, meaning that the learners take responsibility for creating meaning while including the participation of others in confirming worthwhile knowledge. Learning outcomes are perceived as both personally meaningful and socially worthwhile. It is this perspective that makes this model a suitable one to discuss the implementation of self-directed learning in the ATLAS programme. The programme allows students to create a unique academic profile as a New Engineer that aligns with their personal interests and motives. At the same time, the academic institution creates a social environment where their development unfolds. This environment offers ample learning opportunities, but their learning is also socially related.

The social environment has several aspects operating on different levels: national and local rules and regulations, educational missions and visions, the intended learning outcomes of a programme, standards for good quality education and assessment, the programme's rules and regulations, educational procedures and student requirements, organisational structures, the teaching environment, and peers and teachers. All these aspects form a box in which students' learning takes place, meaning that in their personal learning journey, students continuously need to relate to whatever is inside the box. Their learning is never completely free. The fact that there is such a box challenges the opportunity for creating a self-directed learning experience: how can one learn "on one's own" while there are so many factors to relate to, and that may potentially restrict doing so?

4.4.1 Self-Management

Self-management is the first dimension in Garrison's model. Self-management concerns "the enactment of learning goals and the management of learning resources and support" (Garrison, 1997, p. 22). Having choice, resources, support, and feedback opportunities at hand is seminal to self-management. Although it may seem a paradox, self-management in an educational context does not mean learners are left to their own devices—the learning is facilitated through various means, thereby creating a collaborative experience.

How is self-management shaped in the ATLAS programme? The three-year programme comprises six semesters. Each semester has certain objectives or semester goals that all students need to attain. Attaining every goal of each semester means students reach the final learning outcomes of the programme. However, students can attain the semester goals in their own way. This creates the opportunity to build a unique academic profile as a New Engineer. To attain the goals, students propose an academic course list. These courses can be taken at any university in the Netherlands or even abroad. Depending on certain quality standards, online MOOCs (Massive Open Online Courses) can also be on this list. Taken together, there is ample elective choice for students, both in academic content and learning approach. Some university college programmes offer a limited (but still vast) set of choices. In the ATLAS programme, elective choice is limited to whatever university students have access to (bachelor's and master's level courses). In other university colleges that offer a bounded set of choices, the variation in student profiles is predictable. In ATLAS, no student profile will be the same.

To ensure that every student attains the learning outcomes of the programme, the students' elective choices in each semester need to be monitored. Despite being broadly formulated (e.g., "By the end of semester 3, you will be able to critically reflect on the importance of behaving in an ethical and socially responsible manner in mitigating contemporary wicked problems."), courses still need to be aligned with the semester goals. On top of that, certain requirements are imposed. For example, the course credits need to add up to a certain minimum number, and for the later semesters, certain level requirements may apply (e.g., in semester 6, students are expected to pick electives that are on a second or third-year level). If the alignment is there and the requirements are met, the students can begin their semester learning journey. Here, approval by the programme is needed to ensure that students will eventually be able to meet the programme's intended learning goals.

The electives students pick to attain the semester goals offer specific opportunities for learning, support, and feedback from university teachers. The ATLAS programme trusts that the other educational programmes that offer these elective courses are under continuous quality assurance, both with respect to assessment as well as teaching. In the Netherlands, the quality of assessment is monitored through institutional examination boards, and the quality of teaching is safeguarded through continuous teacher professionalisation and entry qualifications.

Students do not enter academia being fully aware of every learning opportunity there is. They might not be aware that fields exist that match their personal interests, motives, and learning needs. To support students in creating a profile that matches their interests, ATLAS's academic advisors help students translate their learning needs into curricular choices. In the ATLAS programme, not only does every teacher have their own domain expertise (on natural science, mathematics, or social science), they are also academic advisors. This means students can ask for curricular advice in different knowledge domains. On top of that, at the end of each semester, students reflect on their learning experiences in the semester to prepare for their elective choices in the next semester. Teachers in ATLAS respond to this reflection and give feedback and advice. Hence, through various means, choice, resources, support, and feedback are provided to the students.

4.4.2 Self-Monitoring

Self-monitoring is the second dimension in Garrison's model (1997). Self-monitoring refers to the process of taking responsibility for personal meaning. To self-monitor one's own learning process means that students ensure that learning accumulates and that learning goals are being met. In an educational setting, self-monitoring depends on both internal and external feedback. The students need to reflect on their learning and evaluate to what extent their personal goals are met, and the programme needs to respond to their reflection and self-evaluation, an illustration of the collaborative constructivist perspective in Garrison's model.

At the start of each semester, all ATLAS students are required to write a Personal Development Plan (PDP). In this plan, they reflect on their current learning needs based on their learning experiences from the previous semester. They express these needs in the context of the build-up of their academic profile as a New Engineer. They are required to align these needs with the objectives each semester, thereby *personalising* their goals. The needs are then translated into curricular choices, leading to a proposed course list that needs approval by the programme. In this process, the inherent tension between institutional requirements and self-directed learning is palpable; not everything goes. The self-directed learning process is shaped within institutional frameworks; hence, it is a collaborative process.

At the end of each semester, students reflect on their learning experiences together with teachers in a meeting called the Self-Evaluation and Reflection (SER). The teachers respond to their reflections to foster reflections that make sense. Students may find it hard to reflect on a deeper level (Bharuthram, 2018), and teachers can help to reach those levels (Guo, 2022). Students are invited to reflect on critical learning incidents (Bruster & Peterson, 2013) and relate these reflections to their learning process of becoming the New Engineer they want to be. Teachers give advice and discuss students' reflections. In this way, personal meaning is addressed and deepened.

Next to reflection, students also self-evaluate their learning at the end of the semester. From the courses they took in their semester, they received feedback and grades. These are inputs for their self-evaluation. Relevant questions for students are: "Did I reach my goals on a sufficient (academic) level? What am I good at, and what is still challenging for me to reach my goals?" Teachers respond to this self-evaluation by asking questions and giving advice. In this collaborative reflection and self-evaluation process, it is assured that self-monitoring is fostered through a process of transactions between internal and external feedback.

Apart from the Personal Development Plan and the Self-Evaluation and Reflection, another aspect of self-monitoring during a semester is the Midterm Evaluation (MTE). In the Midterm Evaluation, students reflect on their learning process and progress halfway through the semester. The Midterm Evaluation is a means to monitor study progress for the programme and is a conversation between the student and teachers. Challenges the students face in their learning process are discussed, and the teachers give tips and advice. The Midterm Evaluation is also a moment to reflect on the curricular choices students made in their Personal Development Plans. Plans can change based on experience, and after the Midterm Evaluation, the students write their Final Personal Development Plan in which they confirm their course list. At this moment, courses can be deleted and added, upon approval of the programme. This (mandatory) learning activity further fosters self-monitoring and allows the programme to track study progress, which is quite challenging to do automatically in a programme in which each student follows their own course.

4.4.3 Motivation

The final dimension in Garrison's model of self-directed learning is motivation. It is beyond the scope of this chapter to discuss all aspects of motivation in relation to self-directed learning, but the most important according to Garrison, is to "create conditions where students become increasingly motivated by authentic interest and desire to construct personal meaning and shared understanding (worthwhile knowledge)" (p. 29). The conditions ATLAS creates to foster motivation are to allow students to build their own study programme and to continuously stimulate reflection on learning and direction without suggesting certain course trajectories.

Anecdotal evidence shows that ATLAS students display high levels of motivation, especially towards the end of the programme. It is believed that, by that time, they have found purpose and are fully aware of their learning needs and capacities to fulfil that purpose. When ATLAS students enter an elective class, they purposefully choose to be there. The course is selected to build up their academic profile, which is theirs. Imagine you are a student who must follow this same class as part of the set of mandatory courses in a more traditional programme. What kind of experience would that be compared to someone who has purposefully and consciously chosen to take that course because it matters as part of one's academic development?

4.5 Motivation and Self-Determination Theory

Although the question previously posed is rhetorical, one can argue, based on the literature on Self-Determination Theory (Ryan & Deci, 2017), that the self-directed learning approach as implemented in the ATLAS programme inherently fosters the motivation to learn. The meta-theory behind SDT states that all human beings want to grow and master challenges. These tendencies need a supportive environment to flourish. However, as Guay (2022) puts it: "in many social contexts, school included, these tendencies are thwarted, leading to a lack of compliance, oppositional behaviours, and disengagement" (p. 75).

Self-Determination Theory distinguishes between three basic psychological needs: autonomy, competence, and relatedness. Autonomy relates to experiencing a sense of choice, willingness, and volition. Competence relates to the experience of being able to interact effectively with one's environment, and relatedness refers to the need for close and secure emotional bonds in collectives (Guay, 2022). The extent to which these needs are fulfilled in an educational environment might help explain the level of motivation students experience in that environment.

With respect to autonomy, ATLAS students cannot complete their education without making choices. The choice options are extensive (students can take electives at every university in the Netherlands or abroad that offers opportunities for exchange), and there are relatively few mandatory components (approximately 40% of the programme is mandatory). On top of that, most of these components (e.g., semester projects) include choice components that allow for pursuing personal goals. Thus, the ATLAS environment affords a high level of autonomy.

With respect to competence, the self-directed learning cycle focuses on evaluating and reflecting on one's learning, together with teachers. This happens in the Personal Development Plan, where students motivate their curricular choices; in the Midterm Evaluation, in which they evaluate and reflect on their study progress; and in the Self-Evaluation and Reflection, in which they look back on the semester and reflect on the consequences of their learning experiences for their future learning. This is done together with teachers, and although the Personal Development Plan, the Midterm Evaluation, and the Self-Evaluation and Reflection are mandatory components of the programme, they are not assessed. Teachers listen to the students' stories about their semester journey and only provide feedback and advice, thereby fostering a safe, low-stakes environment. This also relates to the aspect of relatedness to which we now turn.

Relatedness in ATLAS can be found in many places. For one, the programme is small-scale and residential, meaning that the number of students entering the programme can be overseen. Therefore, students will meet each other daily, both in the classroom and in their leisure time. Second, ATLAS has an active student association that secures onboarding, making sure every student feels safe and at home in the community (in the admission interview, candidates are made aware of the value of community for the programme). All ATLAS students (and many teachers as well) are members of the student association (ATLANTIS) and the vast majority also actively contributes to extracurricular activities the association organises.

Third, being part of a student community also provides a support structure for each student's educational journey. Particularly in making curricular choices, second and third-year students are an important resource for first-year students. Fourth, most of the mandatory components of the programme include collaborative learning. Fifth, the physical environment in which ATLAS resides is designed to foster accessibility and collaboration. Lastly, teachers are close to the students and care about their learning and well-being. In all, these aspects produce an environment that is of the type Guay (2022) argues is needed to support the basic needs discussed.

This analysis, based on Self-Determination Theory, aligns with what is observed in the programme: students taking more courses than the minimum, a high level of extracurricular activity, high grades for bachelor's theses, and high student ratings for the programme.

4.6 Challenges

Implementing the self-directed learning approach in a three-year bachelor's programme also comes with challenges. The ATLAS experience will also be used here to highlight a few aspects that are challenging for students and teachers in the programme.

4.6.1 Onboarding

The first aspect relates to onboarding. First-year students entering ATLAS are informed about the self-directed learning approach, but have seldom or rarely experienced it before. They will compare their new experiences with their previous ones and might get confused. One cannot expect that students will pick up on the approach right away because they chose the programme. Self-directed learning does not work without initial guidance and support. Therefore, first-year students are guided and instructed in managing the specific self-directed learning activities (Personal Development Plan, Midterm Evaluation, Self-Evaluation and Reflection) in the programme. On top of that, every ATLAS teacher serves as a resource for guidance and support.

The onboarding aspect also relates to teachers. New teachers might have specific expectations of what is good education that might be different from what happens in a self-directed programme. They need to be enculturated by experienced peers within the programme to become ATLAS teachers.

It is essential that all aspects of the programme exude the self-directed learning approach, making sure there is no way to go around it, and that both students and teachers learn to work accordingly. This takes time and effort and might explain why implementing self-directed learning in a separate course in the context of a more traditional programme is bound to fail: both students and teachers will lack experience and will conclude that it does not "work."

4.6.2 Study Planning

The second aspect relates to study planning. Being able to pick courses from so many university programmes does not always lead to feasible semester planning. Elective courses might overlap, and students might be forced to let go of a course they want to take because of some mandatory courses that need their attention first.

Typically, the world does not revolve around the ATLAS programme. This is a continuous struggle, forcing students to make difficult choices sometimes. That is why some university colleges offer a fixed set of elective courses that are aligned in such a way that students always have a feasible study plan.

Another aspect of study planning is how to ensure that the students will have a study load that is manageable. ATLAS students are not forced to stick to a minimum number of course credits per semester. They can take more. However, the programme is responsible for safeguarding a feasible study load. The programme struck this balance by requiring students to justify in their Personal Development Plans why they think they can take up a higher load than required. The programme will respond to that, and when the load is considered too high, advice will be given to have a plan B when the load indeed turns out to be unmanageable. However, the programme will never force a student to do less. Students also need to learn what their limits are to become self-directed learners. When they bite off more than they can chew, the programme will warn them and will further support them in making up for lost time and effort.

4.6.3 Fostering Interdisciplinarity

The third challenge is how to foster interdisciplinarity in students. Although not directly related to self-directed learning, interdisciplinarity is one of the five core values of the ATLAS programme. This is expressed in the mandatory courses in mathematics, natural science, and social science, as well as the semester projects that require an interdisciplinary problem-solving approach. These programme components take up 40% of the total programme. However, the remaining 60% is "open." How can the programme motivate students to take courses from different academic disciplines to build an interdisciplinary academic profile as a New Engineer, and not use the elective space to focus on one specific discipline? Students who enter the programme are mostly attracted to the choice options in the programme, so they can be expected to make choices across disciplines. However, during the programme, they might want to focus more on one discipline. This makes sense considering their further (master's) education for which certain entry course requirements might apply. As a result, the level of interdisciplinarity in students' academic profiles becomes a continuum, ranging from monodisciplinary (in the 60%) to truly interdisciplinary. Imposing requirements on the choice of students to foster true interdisciplinarity is an example of the challenge of balancing freedom and confinement.

4.6.4 *The Teacher Team*

A final issue relates to the way of working of the teacher team. One could argue that for developing an interdisciplinary mindset in students, teachers with a similar mindset are required. However, this might not always be the case. University teachers with a PhD are commonly raised in a monodisciplinary environment. To become part of a teacher team for an interdisciplinary programme, they need to relate to teachers from different disciplines in creating and running the programme. This is not an easy task, given a possible lack of mutual understanding of their disciplines and the way teaching is commonly shaped in these disciplines. The same applies to the implementation of a self-directed learning approach. Vangrieken et al. (2015) identified the factors that hinder and foster effective teacher collaboration. Kunnari and Toom (2018) highlight that teams need to find a new mindset for how to shape their work practices collaboratively. A positive and open-minded approach is relevant, as well as an increased awareness of common resources. When starting any new programme, team aspects require careful attention by programme management.

4.7 Conclusion

Shaping a university programme based on self-directed learning principles is a continuous process of striking a balance between freedom and confinement. University education requires adherence to national and local rules and regulations, educational missions and visions, and quality standards for education and educational assessment. These elements create a level of confinement within which the self-directed learning approach needs to take shape. There is an inherent tension here that teacher teams need to consider in any educational design decision to safeguard that the self-directed learning approach will be implemented effectively. With this chapter, I hope to have shown that the question is not *if* it is possible to implement such an approach, but rather *how* this can be done. May this be an inspiration for other universities that see the value of self-directed learning and seek ways to implement it.

References

Bammer, G. (2013). *Disciplining interdisciplinarity: Integration and implementation sciences for researching complex real-world problems*. ANU Press.

Bharuthram, S. (2018). Reflecting on the process of teaching reflection in higher education. *Reflective Practice, 19*(6), 806–817. https://doi.org/10.1080/14623943.2018.1539655

Boyer, S. L., Edmondson, D. R., Artis, A. B., & Fleming, D. (2013). Self-Directed Learning: A Tool for Lifelong Learning: A Tool for Lifelong Learning. *Journal of Marketing Education, 36* (1), 20–32. https://doi.org/10.1177/0273475313494010 (Original work published 2014).

Bruster, B. G., & Peterson, B. R. (2013). Using critical incidents in teaching to promote reflective practice. *Reflective Practice, 14*(2), 170–182. https://doi.org/10.1080/14623943.2012.732945

Garrison, D. R. (1997). Self-directed learning: Toward a comprehensive model. *Adult Education Quarterly, 48*(1), 18–33. https://doi.org/10.1177/074171369704800103

Goldberg, D. A., Sommervile, M., & Whitney, C. (2014). *A whole new engineer: The coming revolution in engineering education*. ThreeJoy Associates.

Guay, F. (2022). Applying self-determination theory to education: Regulations types, psychological needs, and autonomy-supporting behaviors. *Canadian Journal of School Psychology, 37*(1), 75–92. https://doi.org/10.1177/08295735211055355

Guo, L. (2022). How should reflection be supported in higher education?—A meta-analysis of reflection interventions. *Reflective Practice, 23*(1), 118–146. https://doi.org/10.1080/14623943.2021.1995856

Hiemstra, R., & Brockett, R. G. (2012). *Reframing the meaning of self-directed learning: An updated model*. Adult Education Research Conference. https://newprairiepress.org/aerc/2012/papers/22

Kampouroglou, T., Beldad, A., & Wilhelm, P. (2026) Self-directed learning (SDL) after graduation: Graduates' perspectives on its role in shaping their attitudes and behaviours in personal and professional settings. In A. Beldad, P. Wilhelm, J. Homminga (Eds.), *Designing an interdisciplinary engineering education for complex realworld challenges*. Springer.

Knowles, M. S. (1975). *Self-directed learning: A guide for learners and teachers*. Association Press.

Kunnari, I., & Toom, A. (2018). Successful teacher teams in change: The role of collective efficacy and resilience. *International Journal of Teaching and Learning in Higher Education, 30*(1), 111–126.

Loeng, S. (2020). Self-directed learning: A core concept in adult education. *Education Research International, 2020*(1), 1–12. https://doi.org/10.1155/2020/3816132

Morris, T. H. (2019). Self-directed learning: A fundamental competence in a rapidly changing world. *International Review of Education, 65*(4), 633–653. https://doi.org/10.1007/s11159-019-09793-2

Morris, T. H. (2024). Four dimensions of self-directed learning: A fundamental meta-competence in a changing world. *Adult Education Quarterly, 74*(3), 236–254. https://doi.org/10.1177/07417136231217453

Murray, H. (2015). *Lifelong learning in the twenty-first century: An investigation of the interrelationships between self-directed learning and lifelong learning* (Doctoral dissertation, Union Institute and University).

Prince, M. (2004). Does active learning work? A review of the research. *Journal of Engineering Education, 93*(3), 223–231. https://doi.org/10.1002/j.2168-9830.2004.tb00809.x

Ryan, R. M., & Deci, E. L. (2017). *Self-determination theory: Basic psychological needs in motivation, development, and wellness*. The Guilford Press.

University College Twente, Technology, Liberal Arts & Sciences (ATLAS). (2018). *Self-evaluation report 2018: Accreditation document*.

Vangrieken, K., Dochy, F., Raes, E., & Kyndt, E. (2015). Teacher collaboration: A systematic review. *Educational Research Review, 15*, 17–40. https://doi.org/10.1016/j.edurev.2015.04.00

Chapter 5
Let the Social Sciences In: What Matters When Designing Social Science Courses for an Interdisciplinary Engineering Education?

Ardion Beldad

5.1 Introduction

An important impetus for designing and implementing an interdisciplinary engineering education is the recognition among various stakeholders (e.g., academics, policymakers) of the highly complex nature of contemporary challenges that engineers must address. Complex challenges transpiring from the interaction between technologies and people are often depicted as socio-technical (Walker et al., 2008). Adequately tackling socio-technical challenges confronting modern societies, such as the use of artificial intelligence, the energy transition, and digital transformation, requires technical and social solutions (Savaget et al., 2019) that must be sufficiently informed by knowledge from the technical/natural sciences and the social sciences (Kirkegaard et al., 2023).

Despite the benefits of integrating natural science and social science knowledge in addressing wicked problems such as those associated with sustainability, integration efforts are fraught with difficulties, which are caused by uncertainties about integration outcomes, institutional hurdles, and ontological and epistemological differences across disciplines (Persson et al., 2018). Integration challenges also sprout from marked differences in ontological assumptions between the two disciplines, with the natural sciences concentrating on the 'representation of an objective reality' and the social sciences emphasising the construction of an evolving reality (Persson et al., 2018, p. 5).

Although both the natural sciences and the social sciences employ the scientific method to study phenomena (Ledoux, 2002), they differ in approaches used for collecting evidence to answer discipline-specific questions and in the vocabulary

A. Beldad (✉)
University College Twente, University of Twente, Enschede, The Netherlands
e-mail: a.d.beldad@utwente.nl

A. Beldad et al. (eds.), *Designing an Interdisciplinary Engineering Education for Complex Real-World Challenges*, Emergent Directions in Engineering Education 1, https://doi.org/10.1007/978-3-032-20223-9_5

required to make sense of a phenomenon under investigation (Kagan, 2009). The status differential between the two sciences also constrained potential collaborations, with the social sciences adopting defensive strategies to assert its relevance and to counter the public misconception that only the natural sciences possess the solutions to society's problems (Kagan, 2009). Dismissive attitudes towards the social sciences also have an aggravating effect. As Lowe et al. (2013, p. 211) claim, the social sciences have been ridiculed for 'stating the obvious, for hiding behind impenetrable jargon, or for devising grandly universalistic theories and methodologies'.

However, widespread acknowledgement of the view that contemporary challenges society faces cannot be solved through a monodisciplinary standpoint has since withered assertions regarding the natural sciences' monopoly over solutions. Clearly, the social sciences are just as important when tackling modern-day problems, and future engineers who will be instrumental in developing solutions have enough to gain from an education that exposes them to what Kagan (2009) characterises as the two distinct cultures of the natural sciences and the social sciences. Challenges related to sustainability (e.g., water scarcity, biodiversity loss, climate change; Jerneck et al., 2011) and energy demand (Sovacool, 2014a, 2014b), for instance, are recognised as perfect opportunities for social scientists to collaborate with natural scientists in generating much-needed solutions.

In recent years, clamours to integrate the social sciences in engineering curricula (Froyd & Ohland, 2005; Hynes & Swenson, 2013; Sørensen, 2009) have intensified, prodded by the recognition of the vital role the social sciences play in promoting engineering virtues, such as sensitivity to risk, respect for nature, awareness of technologies' social context, and promotion of public good (Harris, 2008). Additionally, training in the social sciences is argued to cultivate engineers' sensitivity to the sociopolitical aspects of their professional environments and to the various institutional considerations for responsible and ethical functioning (Davis, 2012; Zandvoort, 2008). Furthermore, such training helps engineers deal with social science-related questions arising from their activities (Sørensen, 2009).

In response to this clamour for integration, the University of Twente, one of the four Dutch technical universities, established an interdisciplinary engineering education anchored on a liberal arts philosophy. The educational programme emphasises the equal importance of the natural sciences and the social sciences as lenses that students can use to address socio-technical challenges embedded in their educational projects.

In this chapter, the approach used to design and teach mandatory social science courses in an interdisciplinary engineering programme will be discussed. Based on the challenges encountered in teaching those courses, curriculum design-related requirements are identified.

5.2 Embedding the Social Sciences in an Engineering Curriculum

Teaching the social sciences in an engineering programme is predicated on the vision of educating future interdisciplinary engineers who understand how individual, organisational, and societal values shape engineering solutions and their implementation. Conlon (2008) posits that the social sciences help engineers to 'understand the context in which they work and how it both constrains and enables their capacity for social responsibility' (p. 151). Proficiency in the language and ways of thinking in the social sciences allows interdisciplinary engineers to go beyond the traditional functioning of engineering, which has been criticised for neglecting the roles of social forces in realising the effects of generated solutions (Nieusma & Riley, 2010). The centrality of people in the design process (Elliott et al., 2022), therefore, requires engineers to employ a people-centric approach, taking into account how people's values and beliefs impact newly developed solutions (Fila et al., 2014).

Compared to mathematics and the natural sciences, which have long been part of a traditional engineering curriculum (Poulis & Minadakis, 2000; Sheppard et al., 2007; Tryggvason et al., 2001), teaching social science in an engineering programme remains a peripheral pursuit. When taught, social science courses tend to focus on ethical and social questions concerning the undesirable consequences of novel technology adoption (Sørensen, 2009). This narrow focus comes at the expense of other pertinent questions, such as the mechanisms behind the use and appropriation of technology and the roles various social actors play in facilitating the widespread implementation of technical solutions.

A fundamental question concerning the inclusion of the social sciences in an engineering curriculum is the approach taken to realise disciplinary integration. Sørensen (2009) introduces two approaches. The first one prescribes interdisciplinary collaboration among engineers and social scientists when tackling complex real-world challenges, while the second approach involves teaching social science knowledge and skills alongside traditional engineering courses, resulting in a 'hybrid form of engineering knowledge' (p. 93). The second approach is used when teaching the social sciences in our interdisciplinary engineering programme, predicated on our programme's vision to educate a new breed of engineers proficient in the languages of the natural sciences and the social sciences.

5.3 Designing Social Sciences Courses for an Interdisciplinary Engineering Education

Teaching social science courses in our interdisciplinary engineering programme is motivated by three central goals, namely (a) developing and sharpening students' ability to use theories and models in the social sciences when understanding the relationships of various societal actors (e.g., individuals, firms, government institutions) with technologies and the implications of those relationships, (b) introducing students to the different ways of knowledge generation within the social sciences, and (c) training them to communicate newly generated social science knowledge to both academic and non-academic audiences.

Social science courses are mandatory for the first three semesters of the three-year interdisciplinary engineering education. The organisation of each social science course is contingent on the socio-technical challenge selected for the semester. Decisions on theories and concepts introduced in the mandatory social science courses correspond to the levels of analysis in the discipline: micro, meso, and macro. The micro level pertains to interactions at the individual level, with the meso level focusing on exchanges at the group level (e.g., neighbourhoods, communities, formal organisations) and the macro level concentrating on relations at the regional, national, and global levels (Bhandari & Yasunobu, 2009; Reid et al., 2010).

In the first semester of the three-year bachelor's programme, students are expected to develop solutions to challenges related to mobility and well-being. By following the engineering design process described in Chap. 3 by Wilhelm et al.—from problem analysis and idea generation to requirements specification, prototype development, and evaluation—students create a physical artefact that helps individuals with disabilities perform specific tasks. Throughout this project, students use social science as a lens to understand people's decision-making processes in adopting the developed solutions. As Straub (2009) argues, understanding people's technology adoption considerations is crucial since it provides valuable insights into successful technology implementation.

Individuals' technology adoption is a complex process shaped by their perception of the technology. Hence, in the process of facilitating adoption, people's cognitive, affective, and contextual concerns must be understood (Straub, 2009). People's uptake of novel forms of technology is reportedly (e.g., Alaiad & Zhou, 2014; Beldad & Kusumadewi, 2015; Hegner et al., 2019; Huijts et al, 2012) influenced by various factors, such as perception (e.g., risk beliefs), emotion (e.g., anxiety), attitude (e.g., trust), personality characteristics (e.g., innovativeness), and the social environment (e.g., peer influence). The strength of those factors' effects on the adoption process can even vary according to people's cultural values (Srite & Karahanna, 2006; Zhang et al., 2012). Hence, the emphasis on the micro level of analysis is particularly pronounced in this first-semester social science course.

In the second semester, sustainability-related challenges are central. Questions concerning the implementation of and support for sustainable solutions are addressed from the perspective of the social systems theory, which posits that a phenomenon being studied consists of organised and interdependent elements (Holmwood, 2006). When applied in the context of sustainable transition, social system elements refer to the various actors instrumental in realising the sustainable agenda: government institutions, the business sector, non-profit and charitable organisations, the media, academic communities, civic movements, and private consumers/citizens (Koistinen et al., 2020; Lock & Seele, 2017). The multiplicity of social actors related to the challenges of sustainability is unsurprising, as sustainable issues are considered wicked problems (Lönngren & Van Poeck, 2021) involving numerous stakeholders (Peters, 2017).

Just as a social system is inherently complex (Hofman et al., 2017) due to demographic variations, ideological disputes, power struggles and competition, and inequality (Jerneck et al., 2011), sustainability goals are also ambivalent, resulting from various social actors' conflicting interests and values (Newig et al., 2013). The fundamental question for a social science course on sustainability pertains to how interdependence among various social actors shapes the advancement and realisation of the sustainable agenda.

For instance, as Kinzig et al. (2013) reported, environmental challenges can be addressed through modifications in individual actions and the promotion of social norms, which are within the control of consumers (through modifications of consumption patterns) and the government sector (through regulations), respectively. Additionally, growing awareness among consumers of the social and environmental implications of their consumption behaviours for society and the environment has prompted an increase in ethical consumption (Strong, 1997; Wheale & Hinton, 2007).

Consumers who view themselves as socially and environmentally conscious expect companies to behave in ways congruent with consumers' values (Marin et al., 2009). Indeed, while some companies have ethical reasons for embracing corporate social responsibility or CSR (e.g., companies realising that their tasks involve not only production of goods and service but also helping solve social problems partly attributable to their business activities; Buchholz, 1991), firms' CSR initiatives are also viewed as a response to society's expectation of the social and moral responsibilities of businesses (Lindgreen & Swaen, 2010). The focus on various actors in a social science course entitled Social Perspectives on Sustainable Systems completely introduces students to the three levels of analysis—micro (consumers, citizens), meso (commercial organisations, non-profit agencies), and macro (government sector).

In the third semester, the theme of grand challenges requires students to work on a wicked problem, which is characterised as something difficult to define but also has no clear solutions, is essentially unique and possibly a symptom of another problem, involves many stakeholders, and is politically and socially complex (Peters, 2017). Having many stakeholders related to a wicked problem further aggravates difficulties since they often have conflicting interpretations of the problems' emergence (Bannink et al., 2024) and their nature and potential solutions (Head, 2022). Nonetheless, as Bannink et al. (2024) argue, the involvement of various actors in confronting a wicked problem is imperative, as diverse views are needed to fully understand the problem.

A semester focused on wicked problems is an appropriate context for a social science course to engage with governance theories and frameworks, since wicked problems must be 'managed politically' (Head, 2022, p. 38). Moreover, wicked problems can be seriously challenging for policymakers (Termeer et al., 2015), and efforts to tackle an aspect of a wicked problem through public policy involve making choices that trigger ethical questions, specifically when deciding on groups that should benefit from generated solutions and identifying parties who will incur the costs associated with problem-solving (Chan, 2023).

While not officially classified under the social sciences, ethics is a valuable addition to an interdisciplinary engineering curriculum. The gap between social science and ethics has been narrowed by an increase in social science research that expands our understanding of ethical questions (Haimes, 2002). Expectations that engineers will encounter ethical questions when designing solutions, specifically in terms of their accountability for the social impact of their designs (Van de Poel & Verbeek, 2006), are prompting calls for engineering students to be meaningfully exposed to multiple ethical issues (Schuelke-Leech, 2020).

5.4 Requirements for Designing Mandatory Social Science Courses for an Interdisciplinary Engineering Education

5.4.1 Prudent Selection of Concepts to Introduce in the Social Science Courses

Including social science courses in an interdisciplinary engineering curriculum is associated with several issues. A programme with a broad focus must contend with the limited amount of time that could be dedicated to different courses, prompting questions regarding the depth and scope of social science courses being offered.

The broad spectrum of concepts and theories in the social sciences requires careful selection of what should be introduced in the three mandatory social science courses. As previously mentioned, the course contents are aligned with the main themes of the semesters. This corresponds to the requirement of knowledge integration within a semester described in the third chapter of this volume.

An important consideration is that the selected concepts and theories introduce students to typical questions at different levels of analysis in the social sciences. More importantly, the selected concepts reflect the educational programme's desire to develop social science courses valuable for interdisciplinary engineers' task of solving contemporary socio-technical challenges. Figure 5.1 shows the main concepts and sub-concepts across the three levels (micro, meso, macro) of the social sciences. The figure depicts the forces influencing technology use, appropriation, and implementation at different levels – individual, organisational, governmental, and institutional – and the implications of adoption for behavioural, organisational, and societal change, and for the eventual redesign or modification of the technology.

The three courses' emphasis on understanding the dynamic and complex interaction between society and technology narrows their scope and the range of social science theories and models students are introduced to. A common theme among the three social science courses is the focus on complexity, especially since social systems, as Mathews et al. (1999) argue, are unstable and unpredictable.

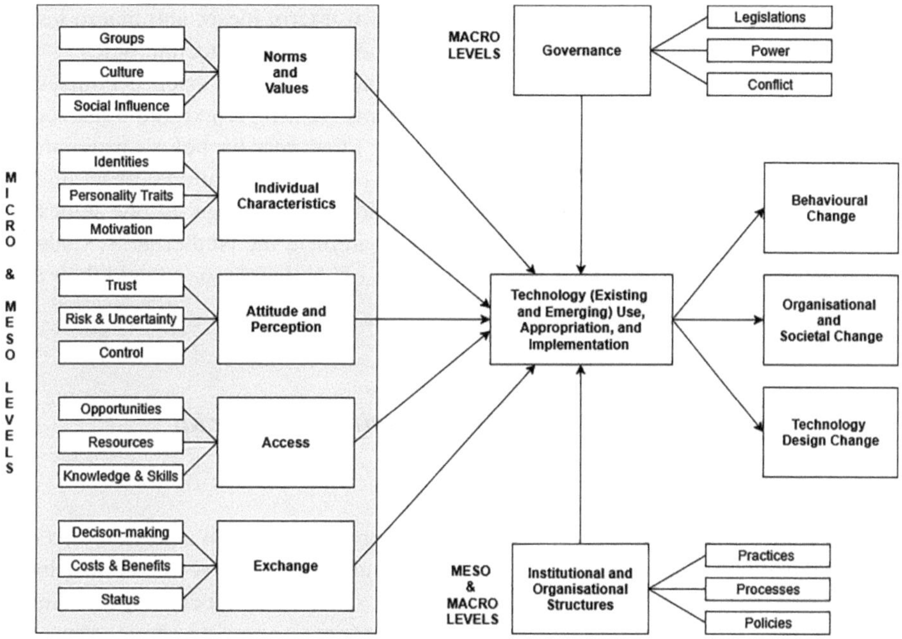

Fig. 5.1 Framework for selecting concepts for the three mandatory social science courses

Additionally, the focus on how various social actors or entities interact, resulting in dynamic behaviour and change within the social system, is expected to sharpen students' systems thinking skills (Arnold & Wade, 2015), especially when assessing how the challenges associated with socio-technical systems and their potential solutions are outcomes of the interrelationships among social system components. Systems thinking, for instance, becomes salient in efforts to realise sustainable transformations, which are shaped by interconnections among individual behaviour, business and economic priorities, and government policies (Voulvoulis et al., 2022).

Five propositions emerge from the framework used to select the concepts for the social science courses. First, at the core of these obligatory courses for an interdisciplinary engineering programme is the interaction between social and technical systems. Second, the courses must help students understand the mechanisms behind technology adoption and implementation at the micro (individual), meso (organisational), and macro (societal) levels, as well as their corresponding implications for individuals, organisations, and society.

Third, understanding the roles of various stakeholders at the micro, meso, and macro levels in the technology diffusion process is a prerequisite when examining the complex dynamics involved in technology adoption and appropriation. Fourth, reflecting on the implications of technology appropriation requires insights into how changes in individual behaviours, organisational strategies, and institutional policies are shaped by the degree of technology appropriation at different levels.

Fifth, ways of technology use and appropriation at micro, meso, and macro levels have important implications for the design and modification of technologies.

As it is unfeasible for the programme's mandatory social science courses to cover a wide range of concepts and topics, students wanting to explore other social science issues have ample space throughout their three-year bachelor's programme to follow elective social science courses offered by other educational programmes within the University of Twente. During their academic exchange for one semester, which is pursued at an academic institution abroad or in the Netherlands, students can also take additional social science courses to either deepen or expand their disciplinary knowledge repository.

5.4.2 De-emphasising Discipline Specificity in Social Science Courses

Unlike a specialised social science educational programme with a curriculum consisting mostly of discipline-specific courses, social science courses in an interdisciplinary engineering programme must transcend disciplinary focus by combining theories from social science subdisciplines to address questions concerning the interaction between the social system and technology across the micro, meso, and macro levels.

A social science course dealing with how individuals interact with technological solutions and the behavioural outcomes of this relationship employs theories at the intersection of disciplines. The Theory of Planned Behaviour (TPB) and Diffusion of Innovation (DOI), for instance, are established lenses used to understand the mechanisms behind individuals' technology adoption, and both theories have their roots in sociology, psychology, and communication science (Taherdoost, 2018). Theories at the intersection of sociology and anthropology, such as the theory of basic values (Schwartz, 2012), are useful for exploring the implications of cultural value differences for novel technology adoption and use.

Considering the need for collaboration among different actors (e.g., citizens, businesses, governments) in solving problems related to sustainability (Nonet et al., 2022), a social science course concentrating on the roles of social systems in advancing the sustainability agenda must equip interdisciplinary engineering students with the required theoretical tools from social psychology, marketing science, policy studies, and public administration. Moreover, a social science course that introduces students to perspectives in governance studies and ethics will benefit efforts to tackle wicked problems.

5.4.3 Designing Learning Activities that Allow Students to Understand Different Ways of Knowledge Generation in the Social Sciences

Social science courses in an interdisciplinary engineering programme must not only aim to build students' disciplinary knowledge but also to develop skills needed for knowledge generation within the discipline. Heterogeneity in the methods, models, and units of analysis used in the social sciences requires students to be aware of their use when studying behavioural and social phenomena. Hence, learning activities organised in social science courses must provide students with opportunities to experience the complete research cycle, which requires in-depth engagement with the literature on a particular behavioural or social problem. This engagement exposes them not only to the conventions of knowledge production in the social sciences but also to the standards for disseminating newly acquired knowledge.

In the first semester social science course entitled Behavioural Science & Technology, for instance, students are required to experience the complete research cycle—from problem identification to data analysis and discussion of results and implications of the research findings —by conducting a quantitative study using either an experimental or a correlational survey method on a specific topic under the broader theme of technology interaction and adoption. The learning activity enables students not only to apply theories introduced in the course and to acquire valuable research skills, but also to develop collaboration and communication skills, while also sharpening a range of metacognitive skills (e.g., analytical and critical thinking, logical reasoning).

As students move to the second-semester social science course, Social Perspectives on Sustainable Systems, they experience a qualitative research approach through a small-scale assignment to explore people's sustainable consumption behaviour. While the activity does not require the delivery of a complete research report, unlike in the first semester, it exposes students to the standard qualitative methods, such as in-depth interviews and focus group discussions.

5.4.4 Providing Students with Learning Experiences that Allow Them to Develop Communication and Collaboration Skills

Research activities and case writing assignments organised in the mandatory social science courses enable students to develop relevant communication skills, such as academic writing (when writing a research report) and oral presentation (when publicly discussing findings of their small-scale research). The fact that research activities are done in small groups of four to five students ensures that communication skills for effective collaboration are also developed. This emphasis on developing interdisciplinary engineering students' communication competencies is a response to expectations in the professional field that engineers can competently engage in dialogues with various social actors (e.g., peers, citizens, policymakers; Ravesteijn et al., 2006).

Instructing students to conduct research in a group is one way to expose them to the benefits (e.g., creative research ideas) and the potential challenges (e.g., problematic communication) of collaborative research (Peterson, 1993). Fostering collaboration in research activities aligns with an interdisciplinary engineering education's emphasis on collaboration as instrumental to effectively tackling complex social problems (Bates et al., 2022; Kokotsaki et al., 2016; Schijf et al., 2023).

5.4.5 Fostering Metacognitive and Problem-Solving Skills Through Various Learning Activities in Social Science Courses

Dramatic economic and societal changes and transformations in business processes, Kay and Greenhill (2010) argue, amplify the urgency to develop students' twenty-first-century skills, such as communicating and collaborating with various individuals across cultural and geographic boundaries, critical thinking, problem solving, entrepreneurial and creative thinking, and knowledge application for creating novel products, processes, and services. Caeiro-Rodríguez et al. (2021) refer to these twenty-first-century skills as soft skills that must be prioritised in an engineering curriculum. The authors cluster these skills into five, namely technical skills (e.g., digital literacy), metacognitive skills (e.g., critical and analytical thinking, integrating and synthesising information), intrapersonal skills (e.g., flexibility, creativity), interpersonal skills (e.g., written and oral communication, collaboration), and problem-solving skills (e.g., explaining problems, project management).

Aside from communication and collaboration skills actively developed through team-based research assignments, learning activities organised throughout the three obligatory social science courses also intend to hone interdisciplinary engineering students' metacognitive and problem-solving skills. For instance, in the course Behavioural Science & Technology, students must individually analyse a real-world case (e.g., individual interaction with a technology) using insights and theories from selected literature.

Completing this assignment requires students to evaluate information from sources, use scientific knowledge to explain a phenomenon, and construct a logically structured story when discussing the case. In the course Social Perspectives on Social Systems, students must complete close reading assignments that require them to demonstrate the ability to critically read selected papers on a topic and reflect on those papers' strengths, weaknesses, and common grounds.

Moreover, the research project in the first-semester social science course is a perfect vehicle for students to develop metacognitive (e.g., critical and analytical thinking, information synthesis and application) and problem-solving (e.g., problem identification and definition) skills.

5.5 Conclusion

Complex challenges confronting modern society require a new breed of problem solvers able to generate solutions resulting from the integration of multidisciplinary expertise. Clearly, a disciplinary-siloed approach to solving a range of complex challenges is no longer effective. What the world increasingly needs are engineers who can comfortably navigate different 'disciplinary cultures' and adeptly work with various social actors, from academics to policymakers to citizens, to develop much-needed solutions. Such a demand necessitates academic institutions to gradually transition from a monodisciplinary approach to education to one that develops students' interdisciplinary competencies.

In recent years, there has been a surge in the number of educational programmes with an interdisciplinary orientation. Acknowledging the urgent need for interdisciplinary engineers, the University of Twente in the Netherlands launched a three-year bachelor's programme designed to educate what it refers to as a new breed of engineers. Hinged on a liberal arts educational philosophy, the programme highlights the equal importance of the social sciences, the natural sciences, and mathematics in its interdisciplinary engineering curriculum. Throughout the educational programme, the social sciences are promoted as highly indispensable for efforts to address complex socio-technical challenges confronting modern society.

An important consideration in the design of social science courses is that the concepts, models, and theories introduced are relevant to the socio-technical challenges students must tackle in the semester projects. However, the narrow focus on specific themes has implications for the depth and the breadth of social science knowledge students will acquire. The open and broad character of the liberal arts programme focused on interdisciplinary engineering signifies that specialisation in the social sciences is a choice that students must make as they progress in the three-year bachelor's programme.

One critical insight to be learned from years of teaching the social sciences in an interdisciplinary engineering programme is that providing students with opportunities to acquire valuable social science skills (e.g., doing research, critical reading of scientific literature) is just as important as exposing them to social science concepts, models, and theories. The challenges of embedding the social sciences in an interdisciplinary engineering are offset by an increasing acknowledgement of the pivotal role the discipline plays in adequately solving complex challenges of our times.

References

Alaiad, A., & Zhou, L. (2014). The determinants of home healthcare robots adoption: An empirical investigation. *International Journal of Medical Informatics, 83(11), 825–840.* https://doi.org/10.1016/j.ijmedinf.2014.07.003

Arnold, R. D., & Wade, J. P. (2015). A definition of systems thinking: A systems approach. *Procedia Computer Science, 44*, 669–678. https://doi.org/10.1016/j.procs.2015.03.050

Bannink, D., Sancino, A., & Sorrentino, M. (2024). Governance without we. Wicked problems and collaborative governance. *Public Policy and Administration, 39*(3), 301–323. https://doi.org/10.1177/09520767241239863

Bates, R., Brenner, B., Schmid, E., Steiner, G., & Vogel, S. (2022). Towards meta–competences in higher education for tackling complex real–world problems—A cross-disciplinary review. *International Journal of Sustainability in Higher Education, 23*(8), 290–308. https://doi.org/10.1108/IJSHE-06-2021-0243

Beldad, A., & Kusumadewi, M. C. (2015). Here's my location, for your information: The impact of trust, benefits, and social influence on location sharing application use among Indonesian university students. *Computers in Human Behaviour, 49, 102–110.* https://doi.org/10.1016/j.chb.2015.02.047

Bhandari, H., & Yasunobu, K. (2009). What is social capital? A comprehensive review of the concept. *Asian Journal of Social Science, 37*(3), 480–510. https://doi.org/10.1163/156853109X436847

Buchholz, R. A. (1991). Corporate responsibility and the good society: From economics to ecology. *Business Horizons, 34*(4), 19–31.

Caeiro-Rodríguez, M., Manso-Vázquez, M., Mikic-Fonte, F. A., Llamas-Nistal, M., Fernández-Iglesias, M. J., Tsalapatas, H., et al. (2021). Teaching soft skills in engineering education: A European perspective. *IEEE Access, 9*, 29222–29242. https://doi.org/10.1109/ACCESS.2021.3059516

Chan, J. K. (2023). The ethics of wicked problems: An exegesis. *Socio-Ecological Practice Research, 5*(1), 35–47. https://doi.org/10.1007/s42532-022-00137-3

Conlon, E. (2008). The new engineer: Between employability and social responsibility. *European Journal of Engineering Education, 33*(2), 151–159. https://doi.org/10.1080/03043790801996371

Davis, M. (2012). "Ain't no one here but us social forces": Constructing the professional responsibility of engineers. *Science and Engineering Ethics, 18*, 13–34. https://doi.org/10.1007/s11948-010-9225-3

Elliott, R., Loh, C. G., Psenka, C. E., Lewis, J. M., Kim, K.-Y., Haapala, K. R., Neal, D., & Okudan Kremer, G. E. (2022). Advancing transformative STEM learning: Converging perspectives from education, social science, mathematics, and engineering. *Journal of Integrated Design & Process Science, 26*(3–4), 393–414. https://doi.org/10.3233/JID-220006

Fila, N. D., Hess, J., Hira, A., Joslyn, C. H., Tolbert, D., & Hynes, M. M. (2014, October). The people part of engineering: Engineering for, with, and as people. In *2014 IEEE Frontiers in Education Conference (FIE) Proceedings* (pp. 1–9). IEEE.

Froyd, J. E., & Ohland, M. W. (2005). Integrated engineering curricula. *Journal of Engineering Education, 94*(1), 147–164. https://doi.org/10.1002/j.2168-9830.2005.tb00835.x

Haimes, E. (2002). What can the social sciences contribute to the study of ethics? Theoretical, empirical and substantive considerations. *Bioethics, 16*(2), 89–113. https://doi.org/10.1111/1467-8519.00273

Harris, C. E. (2008). The good engineer: Giving virtue its due in engineering ethics. *Science and Engineering Ethics, 14*, 153–164. https://doi.org/10.1007/s11948-008-9068-3

Head, B. W. (2022). *Wicked problems in public policy: Understanding and responding to complex challenges*. Palgrave Macmillan.

Hegner, S. M., Beldad, A. D., & Brunswick, G. J. (2019). In automatic we trust: Investigating the impact of trust, control, personality characteristics, and extrinsic and intrinsic motivations on the acceptance of autonomous vehicles. *International Journal of Human–Computer Interaction, 35(19), 1769–1780.* https://doi.org/10.1080/10447318.2019.1572353

Hofman, J. M., Sharma, A., & Watts, D. J. (2017). Prediction and explanation in social systems. *Science, 355*(6324), 486–488. https://doi.org/10.1126/science.aal3856

Holmwood, J. (2006). Social systems theory. In B. S. Turner (Ed.), *The Cambridge dictionary of sociology* (pp. 587–588). Cambridge University Press.

Huijts, N. M., Molin, E. J., & Steg, L. (2012). Psychological factors influencing sustainable energy technology acceptance: A review-based comprehensive framework. *Renewable and Sustainable Energy Reviews, 16(1), 525–531.* https://doi.org/10.1016/j.rser.2011.08.018

Hynes, M., & Swenson, J. (2013). The humanistic side of engineering: Considering social science and humanities dimensions of engineering in education and research. *Journal of Pre-College Engineering Education Research, 3*(2), 1–12. https://doi.org/10.7771/2157-9288.1070

Jerneck, A., Olsson, L., Ness, B., Anderberg, S., Baier, M., Clark, E., et al. (2011). Structuring sustainability science. *Sustainability Science, 6*, 69–82. https://doi.org/10.1007/s11625-010-0117-x

Kagan, J. (2009). *The three cultures: The natural sciences, the social sciences, and the humanities in the 21st century*. Cambridge University Press.

Kay, K., & Greenhill, V. (2010). Twenty-first century students need 21st century skills. In G. Wan & D. Gut (Eds.), *Bringing schools into the 21st century* (pp. 41–65). Springer Netherlands.

Kinzig, A. P., Ehrlich, P. R., Alston, L. J., Arrow, K., Barrett, S., Buchman, T. G., et al. (2013). Social norms and global environmental challenges: The complex interaction of behaviors, values, and policy. *Bioscience, 63*(3), 164–175. https://doi.org/10.1525/bio.2013.63.3.5

Kirkegaard, J. K., Rudolph, D. P., Nyborg, S., Solman, H., Gill, E., Cronin, T., & Hallisey, M. (2023). Tackling grand challenges in wind energy through a socio-technical perspective. *Nature Energy, 8*(7), 655–664. https://doi.org/10.1038/s41560-023-01266-z

Koistinen, K., Teerikangas, S., Mikkilä, M., & Linnanen, L. (2020). Active sustainability actors: A life course approach. *Sustainable Development, 28*(1), 208–223. https://doi.org/10.1002/sd.1989

Kokotsaki, D., Menzies, V., & Wiggins, A. (2016). Project-based learning: A review of the literature. *Improving Schools, 19*(3), 267–277. https://doi.org/10.1177/1365480216659733

Ledoux, S. F. (2002). Defining natural sciences. *Behaviorology Today, 5*(1), 34–36.

Lindgreen, A., & Swaen, V. (2010). Corporate social responsibility. *International Journal of Management Reviews, 12*(1), 1–7. https://doi.org/10.1111/j.1468-2370.2009.00277.x

Lock, I., & Seele, P. (2017). Theorizing stakeholders of sustainability in the digital age. *Sustainability Science, 12*(2), 235–245. https://doi.org/10.1007/s11625-016-0404-2

Lönngren, J., & Van Poeck, K. (2021). Wicked problems: A mapping review of the literature. *International Journal of Sustainable Development and World Ecology, 28*(6), 481–502. https://doi.org/10.1080/13504509.2020.1859415

Lowe, P., Phillipson, J., & Wilkinson, K. (2013). Why social scientists should engage with natural scientists. *Contemporary Social Science, 8*(3), 207–222. https://doi.org/10.1080/21582041.2013.769617

Marin, L., Ruiz, S., & Rubio, A. (2009). The role of identity salience in the effects of corporate social responsibility on consumer behavior. *Journal of Business Ethics, 84*(1), 65–78. https://doi.org/10.1007/s10551-008-9673-8

Mathews, K. M., White, M. C., & Long, R. G. (1999). Why study the complexity sciences in the social sciences? *Human Relations, 52*, 439–462. https://doi.org/10.1177/001872679905200402

Newig, J., Schulz, D., Fischer, D., Hetze, K., Laws, N., Lüdecke, G., & Rieckmann, M. (2013). Communication regarding sustainability: Conceptual perspectives and exploration of societal subsystems. *Sustainability, 5*(7), 2976–2990. https://doi.org/10.3390/su5072976

Nieusma, D., & Riley, D. (2010). Designs on development: Engineering, globalization, and social justice. *Engineering Studies, 2*(1), 29–59. https://doi.org/10.1080/19378621003604748

Nonet, G. A.-H., Gössling, T., Van Tulder, R., & Bryson, J. M. (2022). Multi-stakeholder engagement for the sustainable development goals: Introduction to the special issue. *Journal of Business Ethics, 180*(4), 945–957. https://doi.org/10.1007/s10551-022-05192-0

Persson, J., Hornborg, A., Olsson, L., & Thorén, H. (2018). Toward an alternative dialogue between the social and natural sciences. *Ecology and Society, 23*(4), 1–11. https://www.jstor.org/stable/26796861

Peters, B. G. (2017). What is so wicked about wicked problems? A conceptual analysis and a research program. *Policy and Society, 36*(3), 385–396. https://doi.org/10.1080/14494035.2017.1361633

Peterson, B. J. (1993). The costs and benefits of collaborative research. *Estuaries, 16*(4), 913–918. https://doi.org/10.2307/1352449

Poulis, D., & Minadakis, J. (2000). Designing an engineering curriculum for the 21st century: A case study from Greece. *Industry and Higher Education, 14*(2), 121–124. https://doi.org/10.5367/000000000101294968

Ravesteijn, W., Graaff, E. D., & Kroesen, O. (2006). Engineering the future: The social necessity of communicative engineers. *European Journal of Engineering Education, 31*(1), 63–71. https://doi.org/10.1080/03043790500429005

Reid, L., Sutton, P., & Hunter, C. (2010). Theorizing the meso level: The household as a crucible of pro-environmental behaviour. *Progress in Human Geography, 34*(3), 309–327. https://doi.org/10.1177/0309132509346994

Savaget, P., Geissdoerfer, M., Kharrazi, A., & Evans, S. (2019). The theoretical foundations of sociotechnical systems change for sustainability: A systematic literature review. *Journal of Cleaner Production, 206*, 878–892. https://doi.org/10.1016/j.jclepro.2018.09.208

Schijf, J. E., van der Werf, G. P., & Jansen, E. P. (2023). Measuring interdisciplinary understanding in higher education. *European Journal of Higher Education, 13*(4), 429–447. https://doi.org/10.1080/21568235.2022.2058045

Schuelke-Leech, B. A. (2020, November). The place of wicked problems in engineering problem solving: A proposed taxonomy. In *2020 IEEE International Symposium on Technology and Society (ISTAS)* (pp. 361–372). IEEE.

Schwartz, S. H. (2012). An overview of the Schwartz theory of basic values. *Online Readings in Psychology and Culture, 2*(1), 11. https://doi.org/10.9707/2307-0919.1116

Sheppard, S., Colby, A., Macatangay, K., & Sullivan, W. (2007). What is engineering practice? *International Journal of Engineering Education, 22*(3), 429–438.

Sørensen, K. H. (2009). The role of social science in engineering. In A. Meijers (Ed.), *Philosophy of Technology and Engineering Sciences* (pp. 93–115). North-Holland.

Sovacool, B. K. (2014a). What are we doing here? Analyzing fifteen years of energy scholarship and proposing a social science research agenda. *Energy Research & Social Science, 1*, 1–29. https://doi.org/10.1016/j.erss.2014.02.003

Sovacool, B. K. (2014b). Diversity: Energy studies need social science. *Nature, 511*(7511), 529–530. https://doi.org/10.1038/511529a

Srite, M., & Karahanna, E. (2006). The role of espoused national cultural values in technology acceptance. *MIS Quarterly, 30*, 679–704. https://doi.org/10.2307/25148745

Straub, E. T. (2009). Understanding technology adoption: Theory and future directions for informal learning. *Review of Educational Research, 79*(2), 625–649. https://doi.org/10.3102/0034654308325896

Strong, C. (1997). The problems of translating fair trade principles into consumer purchase behavior. *Marketing Intelligence & Planning, 15*, 32–37. https://doi.org/10.1108/02634509710155642

Taherdoost, H. (2018). A review of technology acceptance and adoption models and theories. *Procedia Manufacturing, 22*, 960–967. https://doi.org/10.1016/j.promfg.2018.03.137

Termeer, C. J., Dewulf, A., Breeman, G., & Stiller, S. J. (2015). Governance capabilities for dealing wisely with wicked problems. *Administration & Society, 47*(6), 680–710. https://doi.org/10.1177/0095399712469195

Tryggvason, G., Thouless, M., Dutta, D., Ceccio, S. L., & Tilbury, D. M. (2001). The new mechanical engineering curriculum at the University of Michigan. *Journal of Engineering Education, 90*(3), 437–444. https://doi.org/10.1002/j.2168-9830.2001.tb00624.x

Van de Poel, I., & Verbeek, P. P. (2006). Ethics and engineering design. *Science, Technology, & Human Values, 31*(3), 223–236. https://doi.org/10.1177/0162243905285838

Voulvoulis, N., Giakoumis, T., Hunt, C., Kioupi, V., Petrou, N., Souliotis, I., & Vaghela, C. J. G. E. C. (2022). Systems thinking as a paradigm shift for sustainability transformation. *Global Environmental Change, 75*, 1–7. https://doi.org/10.1016/j.gloenvcha.2022.102544

Walker, G. H., Stanton, N. A., Salmon, P. M., & Jenkins, D. P. (2008). A review of sociotechnical systems theory: A classic concept for new command and control paradigms. *Theoretical Issues in Ergonomics Science, 9*(6), 479–499. https://doi.org/10.1080/14639220701635470

Wheale, P., & Hinton, D. (2007). Ethical consumers in search of markets. *Business Strategy and the Environment, 16*(4), 302–315. https://doi.org/10.1002/bse.484

Zandvoort, H. (2008). Preparing engineers for social responsibility. *European Journal of Engineering Education, 33*(2), 133–140. https://doi.org/10.1080/03043790802024082

Zhang, L., Zhu, J., & Liu, Q. (2012). A meta-analysis of mobile commerce adoption and the moderating effect of culture. *Computers in Human Behavior, 28*(5), 1902–1911. https://doi.org/10.1016/j.chb.2012.05.008

Chapter 6
Mathematics Across Disciplines: Reflections on Teaching Mathematics at the University College Twente

Georgia Kittou

6.1 Introduction

Extreme inhomogeneities mark the global landscape of our living worlds. On the one hand, our century is characterised by remarkable technological achievements, digital transformation, and unlimited access to information. On the other hand, humanity is exposed to consistent public health crises, social injustice, and climate change. At the same time, modern society promotes social isolation and emphasises individualism, overshadowing collective well-being and mutual support. Is this truly the way forward? Is this the legacy we want to pass on to the next generations? In his Meditation XVII, John Donne wrote the famous phrase "No man is an island" (Donne, 1959), calling individuals to reflect on the illusion of self-efficacy. Today, we need to remind ourselves that human lives, actions, and consequences are interconnected.

This perspective aligns closely with the principles of mathematical modelling, in which solving complex problems requires integration of multiple variables, disciplines, and systems thinking. Mathematical modelling lies its foundations on shared knowledge, collaborative effort, and conceptual understanding. A framework that mirrors the interconnected fabric of society itself. In philosophy and early physics, no model is created in isolation. Instead, models are shaped through interaction. Aristotle's principle that "nature abhors a vacuum" serves not only as a physical claim but also as a metaphor for knowledge creation that supports the growing recognition that "No single discipline has the answer anymore. There is no escape. Read and re-skill like mad" (Herrero, 2021).

G. Kittou (✉)
University College Twente, University of Twente, Enschede, The Netherlands
e-mail: g.kittou@utwente.nl

A. Beldad et al. (eds.), *Designing an Interdisciplinary Engineering Education for Complex Real-World Challenges*, Emergent Directions in Engineering Education 1, https://doi.org/10.1007/978-3-032-20223-9_6

Interdisciplinary approaches are becoming increasingly integrated into the pedagogical framework of modern education. The primary goal is to equip students with the knowledge and viewpoints needed to address the complex problems they may face in both their personal and professional lives. Although the integration of engineering, social sciences, and humanities into multidisciplinary education is gaining ground, mathematics education is frequently restricted to conventional teaching approaches. These approaches usually prioritise procedural fluency, memorisation, and abstraction over relevance and practical application. Considering that mathematics is the cornerstone of multidisciplinary research, the language of links, systems, and patterns (The Quantitative Literacy Design Team, 2001), this divergence is alarming.

Despite its potential, incorporating mathematics meaningfully into interdisciplinary education is not without challenges. Apart from the contextual framework, students often approach mathematics with anxiety or a fixed mindset, creating barriers to success (Boaler, 2016). These attitudes can significantly reduce motivation and limit student engagement. To unlock the full potential of mathematics, educators must adopt approaches that emphasise real-world relevance, encourage collaboration across disciplines, and foster a classroom culture that values curiosity and critical thinking over rote memorisation (AU Department of Education, 2015).

This chapter reflects on the experience of teaching mathematics in an interdisciplinary liberal arts and sciences programme. It explores the pedagogical strategies used to make mathematics accessible and meaningful, the challenges encountered along the way, and the lessons learned from working in a truly integrative educational environment. It argues that mathematics, especially through the lens of modelling, when taught with care and creativity, can empower students not only to understand the world but to change it.

6.2 The Role of Mathematics in Interdisciplinary Education

A compelling reason for embedding mathematics within interdisciplinary education is its vital role in modelling (Karjanto, 2023; Tytler, 2020). Modelling enables students to translate real-world phenomena into mathematical representations and interpret the results in a meaningful context. This not only enhances their mathematical understanding but also develops necessary problem-solving, critical thinking, and higher-order skills, vital for navigating complex global challenges. Apart from its cognitive function, another important dimension of mathematics is its communicative function. Mathematics enables a common ground and serves as a bridge for shared understanding. It is a medium in which individuals from different academic and cultural backgrounds can connect, communicate, and collaborate using the same language (Venegas-Thayer, 2019).

Furthermore, mathematics supports the transfer of concepts across disciplinary boundaries. For instance, the concept of exponential growth is fundamental in biology (e.g., population dynamics), economics (e.g., compound interest), and computer science (e.g., algorithmic complexity). Recognising such cross-cutting ideas helps students perceive mathematics not as an isolated subject but as a connective framework that links diverse subject fields. This perspective is reinforced by Williams et al. (2016), who highlight the importance of developing theoretical frameworks that foster interdisciplinary coherence across mathematics education.

Crucially, it is the teacher who must take the first step in making these connections visible and meaningful. While curricula may promote interdisciplinary approaches, their implementation depends on the teacher's ability to design learning experiences that reveal the relevance of mathematics across contexts. Teachers have the responsibility to help students build transferable knowledge and develop a mindset that values integrative thinking. These intentional pedagogical choices are essential for educators who aim to actively reshape how mathematics is taught in the classroom. Certainly, these efforts are meaningful and fruitful only if the teacher is supported by institutional frameworks: a consistent and aligned curriculum to the academic goals of the programme, a collaborative teacher community, and the autonomy to foster a safe and exploratory environment for both learning and teaching.

Moreover, interdisciplinary mathematics education must balance the tension between exposure to a wide variety of concepts and a thorough understanding of specific areas. As Doig et al. (2019) emphasise, successful interdisciplinary mathematics education depends on maintaining disciplinary integrity while fostering cross-domain connections. For university and college students pursuing their own academic paths, mathematics is not simply a subject to be learned but a fundamental way of thinking. Essentially, modelling should not be taught superficially. Instead, it should be established as a powerful tool for exploration and understanding.

6.3 Pedagogical Approaches and Course Design

The design of mathematics education within interdisciplinary university and college programmes calls for an ongoing and thoughtful balancing approach. While the core curriculum (especially for the first year, including calculus, statistics, linear algebra, and differential equations) remains consistent and often reflects an engineering trajectory, the students engaging with it do not. Each new cohort brings different academic interests, disciplinary goals, and ways of seeing the world. These shifting dynamics demand more than fixed instructional routines. They require educators to rethink how mathematics is presented, understood, and experienced.

The classroom, in this context, becomes a space to plant intellectual seeds. It introduces core concepts and sparks curiosity. But the growth of ideas often happens beyond the classroom. An educational programme needs to create a flexible and meaningful learning environment that encourages this continued exploration and intellectual development. These ongoing efforts reflect the reality of education today: the need to remain grounded in core knowledge while being responsive to the changing world.

6.4 Contextual and Problem-Based Learning

The mathematics curriculum in the ATLAS programme is designed to align with the varying levels of students' prior knowledge and skills. For the first semester, all students enrol in the course Introduction to Mathematical Modelling, which includes a variety of topics from a traditional calculus course, ordinary differential equations, 3-D geometry and vectors, and a range of modelling applications. For the second semester, students are given the option to enrol in elective courses designed in ATLAS to cover multivariable calculus topics and linear algebra (or register in related math courses within UT). MATLAB and DESMOS[1] were also introduced as tools for numerical computation, data visualisation, and modelling. The second- and third-year students have the flexibility to further shape their academic profile in mathematics by selecting from a broad range of elective courses available across the university.

To foster a supportive learning environment for students, especially in the first semester, where students often face challenges transitioning from high school to university, the pedagogical strategies applied are driven by the core principles of problem-based learning. Students are introduced to real-world problems and encouraged to use mathematical modelling to 'translate' the content into mathematical terms. This thematic integration encourages students to draw connections between mathematical concepts and other fields, fostering a more holistic understanding, and encourages the development of cognitive skills such as critical thinking, mathematical reasoning, and problem-solving. Furthermore, this approach encourages students to recognise the importance of mathematics as a tool for understanding real-world phenomena, rather than as an abstract discipline. Furthermore, students take responsibility for their own learning, set and reflect on their learning goals, and emphasis is given to both the learning process and the outcome. The teacher serves as a facilitator rather than a traditional knowledge expert, providing space for students to explore concepts independently, develop teamwork skills, identify their strengths, and acknowledge their weaknesses. This metacognitive component deepens learning and improves future performance.

6.5 In Practice

In Semester 1, the domain course Introduction to Mathematical Modelling is designed to explore how mathematical modelling techniques can be used to analyse and solve real-world problems effectively. Mathematical modelling involves creating abstract representations of real-world systems through mathematical

[1] MATLAB is a programming and numeric computing platform used to analyse data, develop algorithms, and create models.

DESMOS is an open graphic calculator.

expressions and equations. These models simplify complex systems, allowing us to analyse and understand their behaviour in a structured and quantifiable manner. For instance, mathematical models can help us understand population growth patterns, predict the motion of physical systems, or assess the impact of uncertainties in risk management.

One of the key benefits of mathematical modelling is its ability to provide a systematic approach to problem-solving. Through this course, students will learn to develop and apply mathematical models to various scenarios. This process involves identifying key variables, formulating hypotheses, constructing and solving equations, and interpreting results. Students will explore diverse types of models, including those related to population dynamics, mechanical systems, and statistical uncertainties.

In addition to gaining practical skills in differentiation, integration, and differential equations, students will also develop critical thinking and problem-solving skills. The course emphasises an iterative approach to problem-solving: students will formulate problems, develop models, analyse results, and refine their models based on feedback. This approach not only enhances their understanding of mathematical techniques but also prepares them to tackle complex real-world problems with confidence. By the end of the course, students will be well-equipped to apply mathematical modelling techniques to various fields, making informed decisions and contributing to advancements in their areas of interest. The skills acquired in this course will not only aid in academic pursuits but also provide valuable tools for professional and personal problem-solving.

The structure of the course has three main components based on students' engagement, which are self-study and guided study. For a 6 European credit course (14 academic weeks), students are expected to work a total of 12 hours per week. Before the beginning of the semester, the course content (syllabus and meeting times, learning objectives and expectations, study guide and instructions, weekly lessons in the form of presentations and pencasts, recommended exercises, and dates of final assessment) is shared with the students.

A. Preparation (3 hours self-study). This phase focuses on repetition and familiarisation. Each week, students are expected to watch the pencasts, go through each topic slide created, and review the accompanying information. The purpose of these resources is to encourage students to independently recall previously learned facts, build a foundation, and become familiar with new basic concepts through structured input. Understanding is further supported through the recommended exercises, with solved examples provided and explained in the pencasts.
B. Active engagement (6 hours guided study & group work). During this phase, a brief overview of the new topics presented in the pencasts, and slides takes place. Students engage in group discussions, tackle modelling problems, and connect concepts to real-world applications. This is the stage when exploration is conducted through collaborative learning, idea-sharing, and applying theory in new contexts. Additionally, the recommended exercises can be further discussed during tutoring sessions with teaching assistants. The challenge aspect

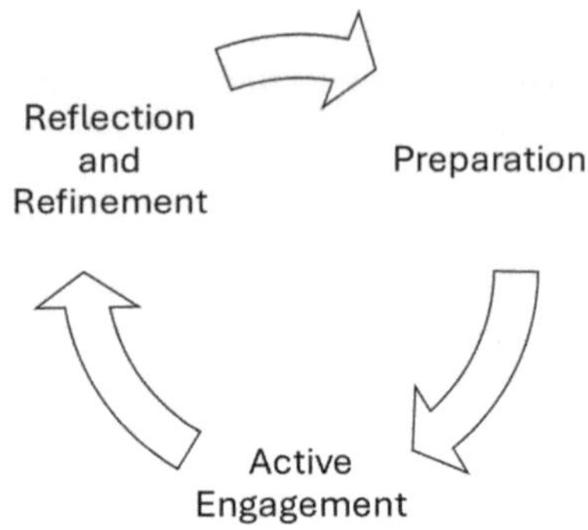

Fig. 6.1 Weekly course flow and student participation

enters when students work on more advanced, application-driven problems that push their understanding.

C. Reflection and refinement (3 hours self-study & group work). This phase leans into exploratory learning again, with reflection and deeper understanding. Students complete the task started in class (in groups, if necessary), submit the work for teacher review, and later engage in a reflection process. This step provides an opportunity for academic growth through feedback and targeted comments (usually done within 2–3 days after the students' submission). After receiving feedback, students participate in a structured reflection process that reinforces learning by encouraging self-assessment. While not as task-intensive as the active phase, it supports learning through metacognition. Figure 6.1 illustrates the cyclical connection among the three phases.

6.6 Syllabus for Elective Courses

In Semester 2, the students have the freedom to choose from a variety of mathematics elective courses offered either at the University College Twente (UCT) or by other faculties within the University of Twente. This allows students to engage with different didactic approaches while ensuring the same mathematical content.

When comparing the courses, the subject content provided is equivalent, ensuring academic consistency. However, the approach in course design, assessment, and teaching follows the framework of UCT.

6.6.1 Multivariable Functions and Optimisation (MATH A)

This course covers the topic of functions of several variables and extends the basic ideas of single variable calculus (domain, limits, continuity, differentiation) to such functions. Multivariable functions are essential to describe physical or real-life phenomena in domains such as epidemiology, population biology and ecology, thermodynamics, electromagnetism, and economics.

6.6.2 Multiple Integration (MATH B)

This course is an introduction to the (Riemann) integration of functions of more than one variable, with an emphasis on integration of functions with two or three variables. After the definition of the integral and the application to integration over simple regions, we consider more complicated areas, curvilinear coordinates, and the transformation of variables.

6.6.3 Vector Spaces and Planar Systems (MATH C)

Building on the intuitive understanding and calculation techniques from Introduction to Mathematical Modelling, this module introduces the concepts of vector spaces and linear maps. Matrices are revisited as the representation of a linear map. In the notion of a subspace, the fundamental concept of a basis for a subspace and the definition of dimension will be introduced. We furthermore introduce the concept of eigenvalues, eigenvectors, and methods to tackle systems of differential equations. We conclude this module by discussing the behaviour of planar systems.

During those elective courses, students are assessed both individually and in groups, and in some cases are asked to perform an experiment as part of the modelling process. For their final challenge, students were asked to choose a topic of their interest (from a list of real-world problems). Topics included multivariable calculus and revenue, optimisation in blood types, can packing, and competition dynamics.

6.7 Interdisciplinary Topics, Diverse Students, and Assessment

As Boaler (2016) underscores, students are more likely to engage with mathematics when they perceive it as meaningful and connected to their everyday lives. This principle guides the design of both in-class activities and final assessments. The flipped-classroom activity ‘Teacher for a Day’ is a pillar of the course flow. This activity empowers students to take ownership of their learning by stepping into the role of the teacher and experiencing the teaching and learning cycle. Students may choose a subject they find interesting from a list of topics or suggest one that aligns with the broader course objectives and deepens understanding (Drilling and Exploratory phase). As teachers, they must prepare a mini-interactive session to present the topic to their peers. After delivering the session, students receive feedback from both peers and the teacher. This process guides students in evaluating their teaching experience, communication skills, and content mastery.

In addition to this, each week, students explore different topics that often encourage learning beyond the classroom environment. For example, in the first semester, topics included conducting experiments to align observations with the theoretical framework of gravitational acceleration, modelling Newton's cooling law, urban design using orthogonal trajectories, and describing the plot of Romeo and Juliet through differential equations based on emotionally written descriptions. Other activities involved error propagation analysis using real marathon athlete data, optimisation problems related to sustainability, and in linear algebra, applying systems of linear differential equations to model group dynamics, exploring eigenvalues in blood type classification, and studying prey-predator models within ecosystems. These projects not only develop mathematical skills but also enhance collaboration with peers, manage team dynamics and conflicts, and communicate ideas. Students learn firsthand to use mathematical modelling to interpret context, not just within a technical framework, but also to highlight their relevance to societal issues.

Each student receives individualised written feedback using the form of WISE feedback (Yeager, 2024) - what went well, improvements, suggestions, and encouragement. This form of detailed feedback also focuses on student skills 'development (structure of document, justification of solutions, arguments used, communicating findings, collaboration, relevance with real-world applications, validity of results, time planning). A detailed rubric is provided by the teacher, outlining performance against technical-specific criteria to ensure transparency and consistency. Students use this feedback and the rubric to reflect on their learning, identify next steps, and set goals for future improvement (Yeager, 2024). If necessary, the student is appointed to further study exercises or content to strengthen their understanding of the topic and gain clarity as part of tailored guidance.

6.8 An Activity Integration Timeline

An applied example is illustrated below to describe the timeframe and components of *a blended learning cycle*, consisting of different learning components, such as experimental exploration, theoretical modelling, and reflective assessment. The different stages of the timeframe are listed below for the group challenge *To Fall or Not to Fall*[2]*;* Modelling object in Free Fall with Air Resistance.

[2] *I try to design my courses in a way that sparks curiosity from the very start. Many science communicators and authors, also use clever, unexpected titles to draw readers into complex topics. A cleverly named task not only adds a touch of personality to the course but also motivates students to engage and invites them to explore further. It's a small touch that can make a big difference in framing the learning experience.*

6.8.1 *Week 1: Experimental Exploration (Discovery Phase)*

The teacher creates a small group of students (after a random selection) and assigns them the task of designing and conducting an experiment to estimate the gravitational acceleration constant g. Furthermore, students are given minimal instructions (e.g., bring appropriate equipment) and asked to connect their experimental process to a theoretical framework of their choosing. This open-ended task encouraged inquiry, collaboration, and critical observation from the start of the course.

6.8.2 *Weeks 2–8: Theoretical Development (Definition & Development Phases)*

As part of the regular course flow, students are introduced to differentiation and integration, both conceptually and computationally. They are guided to frame the problem of free fall with air resistance, derive relevant mathematical models, and use both analytical methods (solving ODEs) and numerical approaches (Euler's method). During this period, students are supported step by step with the conceptual and mathematical tools needed to model and analyse falling motion.

6.8.3 *End of Unit: Assessment & Reflection (Delivery Phase)*

In the unit test and final project deliverables, students are expected to be able to derive the model using Newton's second law of motion (as a starting point), solve the linear differential equations analytically and use numerical methods for the non-linear case. Compare numerical simulations with analytical solutions, connect their findings back to their initial experimental results, and critically reflect on model assumptions, limitations, and potential improvements.

6.9 Pedagogical Value of the Blended Learning Cycle

This process is engaging students in an iterative, authentic scientific inquiry, woven into the course over time. It allows students to explore, define, and validate models based on real-world data. It encourages interdisciplinary thinking, combining physics, calculus, data analysis, and computational skills. And most importantly, students learn to work as a team and as individuals, embrace uncertainty, and appreciate the iterative nature of scientific modelling. The framework described above can be perfectly summarised by the *Double Diamond* process, presented in Fig. 6.2.

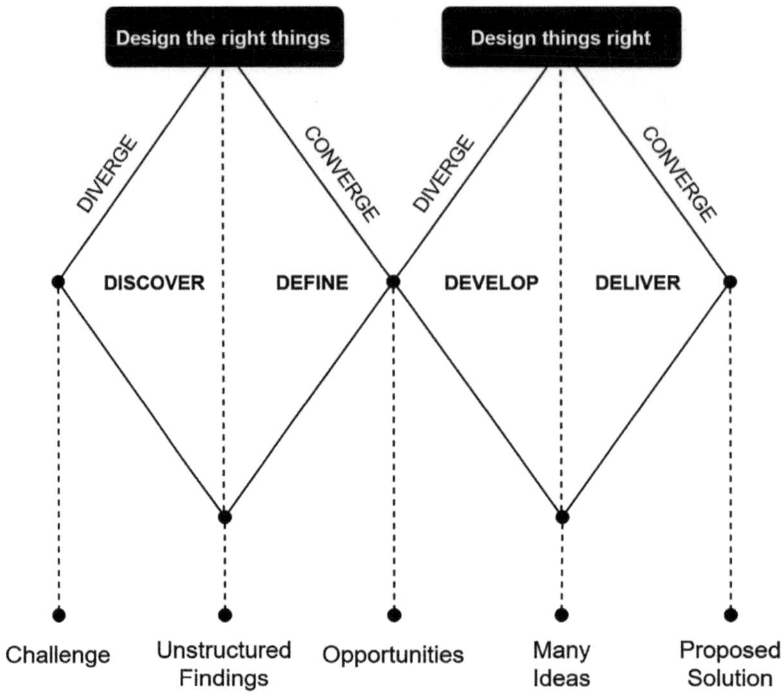

Fig. 6.2 Double Diamond Design Thinking Process for in-class activities as introduced by the Design Council in 2003 (Design Council, n.d.)

6.10 Inclusivity Relative to Scalability

One of the most important aspects of teaching mathematics at University College Twente is the small-scale classroom size of students. Smaller sizes foster a more interactive and engaging environment, enhance student-teacher interactions, create an opportunity for personalised attention, and group collaboration becomes more manageable. The small size of the student group gives both space and capacity to teachers to provide personalised feedback to students, develop a deeper understanding of the individual strengths of each student, and adapt lessons accordingly when necessary. Furthermore, the teacher has the flexibility to employ different teaching methods. Another important feature is *community,* another core idea of the programme. Effective learning takes place when students feel safe and possess a sense of belonging in an inclusive learning environment where diversity is respected and celebrated.

Most students enter university with the belief that they are not good at math, due to prior experiences and exposure to instructional strategies focused on memorisation. These observed students' behaviour leads to disengagement, lack of motivation, insecurity, and math anxiety. That said, in a small classroom environment, these challenges can be more effectively addressed and further developed into opportunities for growth. Consequently, students are more likely to rebuild their mathematical identity, develop conceptual understanding, and gain confidence in their abilities.

6.11 Emerging Challenges of Teaching Math in an Interdisciplinary Engineering Programme

The main challenge is that the implementation of small classrooms within institutions is usually not the preferred approach due to high costs and limited teachers 'capacity. In higher education, scalability and mass delivery models are often prioritised over personalised learning. Establishing more inclusive learning environments requires a cultural transition in conceptualisation and implementation. Nevertheless, components of the University College Programme educational model can still be scaled through creative means, such as hybrid teaching methods, flipped classrooms, and the use of technology to support individualised feedback and engagement. These adaptations offer a way to preserve the spirit of personalised learning, even within the broader logistical constraints of higher education.

Another significant challenge from the students' perspective is that after experiencing the teaching and learning style within the college programme, they find it difficult to fit into a traditional university classroom. And the struggles are not related to the lack of understanding of the course content, but rather to the differences in the course expectations and the way of assessment. The college students are choosing a personalised learning academic path and often feel disoriented when faced with conventional lectures and standardised exams. Despite this shift, students often excel academically, drawing on their strong foundation in self-directed learning, critical thinking, and adaptability, highlighting the strengths developed through the college educational model. Students have the role of trailblazers who are well-equipped to thrive across diverse academic environments and real-world challenges.

Persistent tension in teaching mathematics in an interdisciplinary environment is balancing two key aspects. First is the development of a solid understanding of core mathematical concepts and methods. Second, the curriculum must be accessible and relevant to all students, especially those coming from non-mathematical backgrounds, without overwhelming them. Striking this balance is a significant challenge, as it requires teaching that is both rigorous and inclusive. This challenge is often further complicated since collaboration among the college teachers from diverse disciplines is involved. Communicating and aligning expectations of domain-specific courses to support the project and the semester theme can be difficult, particularly across disciplines that do not traditionally overlap. Addressing this requires ongoing dialogue, flexibility, and a shared commitment to integrated learning outcomes. Despite these challenges, the interdisciplinary approach strengthens the educational experience by promoting holistic thinking and meaningful learning experiences.

6.12 Conclusion

Teaching mathematics in an interdisciplinary university college programme provides academics with the opportunity to learn and grow alongside students and colleagues from diverse academic backgrounds. The role offers learning space and freedom to redesign mathematics education not as a rigid and isolated discipline, but as a dynamic and integrative way of thinking. This holistic approach is pivotal in inspiring and motivating students *to see* the real beauty of mathematics.

Key features that support interdisciplinary education and mathematics didactics are embedded within the educational framework of the educational programme, self-directed learning, integration, excellence, community, and trailblazing. The programme fosters a learning environment that invites students to *wonder and wander*. In parallel to this approach, students are encouraged to shape their own academic and professional paths, establishing autonomy and purpose in their learning journey. By emphasising the societal impact of complex and multidimensional challenges, the programme naturally integrates mathematical modelling into education.

Holistic education recognises that learning is a dynamic process and needs to be initiated, supported, and respected. Under this perspective, no one learns in isolation; education thrives when collaboration, dialogue, inclusivity, diversity, and shared growth are centralised. The academic institution serves as a microcosm of the real world, where diverse perspectives converge, challenges mirror societal complexity, and students are empowered to forge their own paths with purpose and resilience.

References

Australian Academy of Science. (2015). *Desktop review of mathematics school education pedagogical approaches and learning resources*. https://www.education.gov.au/download/2917/desktop-review-mathematics-school-education-pedagogical-approaches-and-learning-resources-june-2015/4043/document/pdf

Boaler, J. (2016). *Mathematical mindsets: Unleashing students' potential through creative math, inspiring messages and innovative teaching*. Jossey-Bass/Wiley.

Design Council. (n.d.). *The double diamond*. Design Council. https://www.designcouncil.org.uk/our-resources/the-double-diamond/

Doig, B., Cheeseman, J., & Roche, A. (2019). Mathematics in the interdisciplinary classroom: Balancing depth and breadth. In R. Barwell & L. van Zoest (Eds.), *Interdisciplinary mathematics education: Curriculum, pedagogy, and research* (pp. 45–63). Springer.

Donne, J. (1959). *Devotions upon emergent occasions: Together with death's duel*. University of Michigan Press.

Herrero, L. (2021, September 6). *No single discipline has the answer anymore. There is no escape. Read and re-skill like mad*. Leandro Herrero Institute. https://leandroherrero.com/no-single-discipline-has-the-answer-anymore-there-is-no-escape-read-and-re-skill-like-mad-2/

Karjanto, N. (2023). *Teaching mathematical modeling for sustainability: Enhancing interdisciplinary skills in students*. Cornell University. arXiv preprint arXiv:2312.02165. https://arxiv.org/abs/2312.02165.

The Quantitative Literacy Design Team. (2001). The case for quantitative literacy. In L. A. Steen (Ed.), *Mathematics and democracy: The case for quantitative literacy* (pp. 1–22). The Woodrow Wilson National Fellowship Foundation.

Tytler, R. (2020). STEM education for the twenty-first century. In J. Anderson & Y. Li (Eds.), *Integrated approaches to STEM education* (pp. 21–43). Springer.

Venegas-Thayer, M. A. (2019). Integration from a commognitive perspective: An experience with mathematics and music students. In B. Doig, J. Williams, D. Swanson, R. Borromeo Ferri, & P. Drake (Eds.), *Interdisciplinary mathematics education* (pp. 79–97). Springer.

Williams, J., Roth, W-M., Swanson, D., & Doig, B. (2016). Interdisciplinary Mathematics Education: A State of the Art. In: Interdisciplinary Mathematics Education. ICME-13 Topical Surveys. Springer, Cham. https://doi.org/10.1007/978-3-319-42267-1_1 https://doi.org/10.1007/978-3-319-42267-1_1

Yeager, D. (2024). *10 to 25: The science of motivating young people: A groundbreaking approach to leading the next generation—And making your own life easier*. Simon & Schuster.

Chapter 7
A Concept-Based Framework for the Natural Science Curriculum in an Interdisciplinary Programme

Martin van der Hoef, Sander Wildeman, and Jasper Homminga

7.1 Introduction

The most pressing and interesting current engineering challenges are not just technical but also have a social component. That is, solving such challenges requires an integrated socio-technical perspective and an understanding of how technical solutions function in the real world, and engineers who can identify connections across boundaries between disciplines and see the interrelatedness of problems and solutions across different fields (Goldberg et al., 2014). Such engineers are particularly needed in helping to address what have become known as Grand Challenges, such as the UN's Sustainable Development Goals (United Nations Development Programme, n.d.).

The aim of the ATLAS programme at the University of Twente is to educate this new kind of engineer. To this end, students work together on semester-long projects that revolve around socio-technical challenges, where they learn to collaborate and integrate knowledge from different domains. A second aspect of the programme is that there is much attention to the individual academic development of students. Aside from following a limited number of required domain courses in the first two years, students largely create their own curriculum.

One of the main challenges in the design of such an engineering-oriented interdisciplinary programme is the selection of required knowledge and skills that form the academic foundation on which the student with an interdisciplinary interest can

M. van der Hoef (✉) · J. Homminga
University College Twente, University of Twente, Enschede, The Netherlands

Faculty of Science and Technology, University of Twente, Enschede, The Netherlands
e-mail: m.a.vanderhoef@utwente.nl

S. Wildeman
University College Twente, University of Twente, Enschede, The Netherlands

A. Beldad et al. (eds.), *Designing an Interdisciplinary Engineering Education for Complex Real-World Challenges*, Emergent Directions in Engineering Education 1, https://doi.org/10.1007/978-3-032-20223-9_7

build, for instance, which set of topics and courses is compulsory for all students in the programme? In this chapter, we focus on this challenge for the Natural Science domain, based on our experience at the University College Twente. We suggest a new framework, where the curriculum is built on concepts rather than disciplines, which was inspired by the works of Stern et al. (2017) and Erickson and Lanning (2013) and the concept-driven curriculum of the International Baccalaureate high-school programme (Medwell et al., 2019).

This chapter is organised as follows: in the next section, we address the challenges that are faced when building a natural science curriculum for an interdisciplinary programme. In Sect. 7.3, we introduce a new setup, where the curriculum is based on four concepts. In Sect. 7.4, we show that this set-up naturally links to elements of the modelling cycle that are commonly used to address problems in natural science. In Sect. 7.5, we propose a possible set-up of the programme over the first four semesters, based on these concepts. We end this chapter with a conclusion and a brief discussion.

7.2 Natural Sciences in an Interdisciplinary Programme

The natural sciences are typically divided into those studying living systems, the life sciences (e.g., medicine, biology), and those studying non-living systems, the physical sciences (e.g., physics, chemistry, earth science). Of these, physics could be considered as the most fundamental, as the other sciences build upon the laws of physics for specific application areas. A typical aspect of the natural sciences is that there is a certain hierarchy, meaning that certain topics have to be studied in a specific order. For example, to understand the reaction of molecules, some background in Quantum Mechanics is required, which in turn requires concepts from Newtonian Mechanics. In the curriculum of an interdisciplinary programme where students also have ample free space to shape their own academic profile, there is not enough 'room' to provide a meaningful scaffolding across the breadth of the natural sciences, and therefore, choices in range and depth have to be made. The question that we want to address in this chapter is what the guiding principles can be for such choices. Since the main goal of our programme is to develop a socio-technical perspective of real-world problems, the foundation in the Natural Sciences should abide by a few conditions:

1. connect to topics from current Grand Challenges, and more specifically, connect to the interdisciplinary social-technical semester projects that are the backbone of the programme, and
2. be of sufficient relevance and interest to those students who continue to develop a profile that is more geared towards the social sciences.

A practical aspect is that many students will follow a master's programme in natural science after ATLAS, and to this end, use part of the elective space to prepare for this. Also, a specific aim of the programme is to guide students in this process. For these reasons, the curriculum in the first two years of the program should also:

3. provide a sufficiently broad perspective of the natural sciences to help student in making a choice for their future academic profile, and
4. provide sufficient foundation to build a curriculum that would allow a student to continue in a master's programme in the natural or engineering sciences.

The Natural Science curriculum that was in place in ATLAS for the last ten years was mainly designed based on condition 1 (connect to the interdisciplinary social-technical semester projects) and condition 4 (provide sufficient foundation for a master's in the natural sciences). With respect to the latter, an analysis of courses that are required for entering a range of natural science master's programmes revealed that Newtonian Mechanics, Thermodynamics, and Transport Phenomena are the most common. We took those three as our basis and added a choice of either Material Science or Electronics, as these two areas have seen rapid developments in the past years and play a role in many of today's technologies. Newtonian Mechanics and Thermodynamics & Transport Phenomena also fitted quite well into the projects of semester 1 (Human Motion) and semester 2 (Sustainable Systems), and have remained the natural science basis of the first-year curriculum.

The choice for the second-year courses has become more guided by condition 3 (provide a sufficiently broad perspective of the natural sciences) in combination with the project theme, and has changed over the years. Condition 2 (be of sufficient relevance for students who focus on the social sciences) was found to be the most challenging and was addressed mainly within the courses themselves, where special attention was paid to what we call beta-literacy. By this, we mean that students become aware of the more general concepts that are typical of natural science. For example, a differential equation is used to describe phenomena or that phenomena are transferable across domains and can borrow insights from each other.

Additionally, students learn to solve problems in natural science via the mathematical modelling cycle, where many of the above-mentioned concepts are naturally addressed. The main purpose of creating this literacy is that a graduate of the programme with a clear profile in social science would be able to understand the language and way of thinking of natural scientists, which would make much more effective communication possible when collaborating in a multidisciplinary team. Still, this beta literacy was addressed in a rather ad-hoc fashion in the courses, often depending on the background and experience of a specific teacher, and not based on a consistent and comprehensive framework.

In this chapter, we introduce a new design of the natural science curriculum, based on our experience over the past ten years, in which the concepts of natural science are leading in the development of the curriculum, rather than the classical disciplines, which only come in as examples of explaining these concepts. There are several advantages of this approach.

- Condition 1 can be easily achieved by putting more focus or depth on a particular topic that would link directly to the project. This can readily change from year to year without disturbing the overall setup.
- Both conditions 2 and 3 are naturally achieved, since beta literacy is now the main goal, and a cross-section of the natural science domain will be used to illustrate the concepts.
- The concepts connect more naturally to elements from the mathematics domain, as we will show in the implementation per semester.
- Some concepts have a clear counterpart in the social sciences, and addressing the equivalences and differences connects well with the interdisciplinary mindset of the programme.

The setup will be least favourable with respect to condition 4 (preparing for master's programs in natural sciences), if alone, because the courses may not be identifiable in terms of traditional domain topics such as Thermodynamics. In principle, students should be able to acquire the same basis in natural science that they can acquire through traditional courses; however, we have not yet encountered a setup like this for natural science bachelor's programmes.

7.3 Concepts of Natural Science

The idea to develop a curriculum based on concepts was inspired by the first chapter of the book by Stern et al. (2017) on the principles of concept-based curriculum design, which, in turn, is based on the work by Erickson and Lanning (2013). The main idea is that when you want to apply certain knowledge to a new situation, you need an understanding at a conceptual level. Students will need to learn in terms of concepts rather than just facts and topics. In general, a 'concept' could be considered as an abstract, universal idea or principle that is at the basis of more specific occurrences or examples (Erickson & Lanning, 2013). Erickson and Lanning distinguish macro-concepts from micro-concepts. Macro-concepts are very broad and apply to different disciplines, while micro-concepts underlie the macro-concepts and are more specific to a discipline.

In our approach, we refer to macro-concepts as those that apply to the natural sciences in general, and micro-concepts as the sub-elements that are examples or representations of a macro-concept but are not necessarily specific to a (sub)domain. Micro-concepts are meant to make it more tangible which topics from the natural sciences could be used to connect to a macro-concept (for example, for the macro-concept 'systems', a micro-concept could be 'free-body diagrams', which is more easily recognised as an element that could be taught in a course). Firstly, we collected all concepts (macro or micro) that came from experience with teaching natural science courses and guiding students in multidisciplinary projects. Secondly, we identified how these concepts could be connected (belong to a similar group or could be identified as a sub-concept of the other), which finally brought us to four macro-concepts that we could use as the basis of our education: systems thinking, mathematization of phenomena, origin and hierarchy of laws, and energy. We will describe the concepts in more detail below.

7.3.1 Systems Thinking

In systems thinking, you make specific what elements are part of 'the system', and which are part of 'the surroundings'. The interaction of the system with its surroundings can be seen as input and output, or as a prescribed, constant interaction (boundary conditions). A system can contain subsystems, where the larger system is then the surrounding. The boundaries specifying what defines the system can be with respect to space, properties, or effects to be included.

Examples of a systems thinking approach are the free body diagrams in mechanics (where a body is the system and external forces represent the surrounding), ensembles in thermodynamics (e.g., an NPT ensemble is where the surrounding is a reservoir which prescribes pressure and temperature of a system), and control volumes in transport theory (change of a quantity per unit time is the difference between inflow and outflow). The concept of systems thinking is closely linked to the modelling cycle, in which you aim to simplify a problem so that it becomes manageable and transparent (the 'model'). This form of system thinking is also applicable in the social sciences, where one would, for example, consider a person, while their surroundings are only taken as input/output, or a country where the world is input/output.

As typical *micro-concepts,* we can identify

- *control volumes* in fluid mechanics and process engineering;
- *free-body diagrams* in biomechanics and structural engineering;
- *ensemble theory* in thermodynamics and statistical physics; and
- *transfer functions* in electrical engineering and control engineering

7.3.2 Mathematisation of Phenomena

In natural science, most phenomena are formulated in terms of domain-specific laws that use the language of mathematics. Often, these laws take the form of a simple algebraic expression or a differential equation (For example, Newton's law for mechanical motion, $F = ma$, or Fourier's law for heat flow, $q = -k \nabla T$, or the ideal gas law, PV = nRT). This 'mathematisation' allows one to use mathematics to reason about natural phenomena.

Furthermore, it turns out that when domain-specific laws are combined to model a specific phenomenon (e.g., sound, described using Newton's laws and the ideal gas law), one often finds that the resulting abstract mathematical form (e.g., a 'wave equation') also describes phenomena in a completely different domain (e.g., light, described using Maxwell's equations). Since the same equation has the same solution, this mathematical abstraction facilitates the *transfer* of reasoning about phenomena across the various domains (e.g., the description of reflection and refraction of waves is similar for sound and for light). The concept of mathematisation is less common in the social sciences, where one would work with logical and statistical connections, but much less with mathematical connections.

As *micro-concepts,* we can identify phenomena and/or concepts that are transferable across domains, such as

- *waves* (transport of the disturbance of a quantity)
- *diffusion* (transport of a quantity by random motion of elementary carriers)
- *potential* (a difference in potential drives a flow of some quantity)
- *impedance* (opposition to the flow of some quantity)
- *inertia* (opposition to a *change* in the flow/motion of some quantity)

In this, 'quantity' can be energy (heat), molecules, charge, cells, or matter in general. Examples of these microconcepts are found in almost any area of natural science, such as optics, acoustics, fluids, electronics, materials science, biology, oceanography, and astronomy.

7.3.3 Origin and Hierarchy of Laws

Whereas the previous concept addresses the 'horizontal' relations in the natural sciences, this concept addresses the 'vertical' relations. Where do the laws of nature come from, and how do they build upon each other? Most laws are ultimately empirical in nature: they are a formulation of a general mathematical pattern found to capture a wide range of observational and/or experimental data (as an example, experiments on the relation between pressure P, volume V, and temperature T of a gas led to the ideal gas law $P \bullet V/T = constant$).

Measurement and advancement in measurement technology play an important role in the discovery and formulation of laws. Since measurements always have a limited accuracy, precision, and context, laws are always provisional and approximate to a certain extent. Furthermore, in natural science, researchers have often succeeded in linking known laws to phenomena at smaller scales (giving a deeper understanding of the laws) and to phenomena at higher scales by looking at the collective effect (which may give rise to new laws that have not yet been discovered). As an example, the ideal gas law can be derived from an underlying picture of molecules moving around according to the laws of (quantum) mechanics.

To summarise, the macro concept of the origin and hierarchy of laws is that the body of knowledge of natural science can be seen as a framework where one law underlies the other, where each law is valid or relevant for a particular scale or context, and where, in most cases, the laws are of an empirical origin.

As *micro-concepts*, we can identify examples of laws that could be used to convey the various aspects of this macro concept:

- *Kepler's laws of planetary motion*: Early in the seventeenth century, Kepler formulated his laws of planetary motion based on the mathematical patterns he observed in the accurate astronomical data of Tycho Brahe. Especially, the third law is interesting, as in its formulation, Kepler was inspired by the mathematical relations that govern the sound from vibrating strings (a good example of the role of mathematisation in the natural sciences). Later, Newton could explain Kepler's laws in terms of more fundamental laws of motion (inferred from mathematical patterns in earth-bound motion experiments of, for example, Galileo Galilei and Christiaan Huygens) and his law for the force of gravity.
- *Ohm's law relating voltage and current*: Inspired by the experimental and mathematical work of Fourier on heat transfer (Fourier's law), Ohm (around 1825) sought for similar mathematical patterns in experiments with 'galvanic circuits' (varying the voltage and wire length and observing the current). The formulation of this law was greatly facilitated by new technology to measure current (the galvanometer) and generate a constant voltage (batteries and thermocouples). Scientists, such as Drude and Sommerfield, could later explain Ohm's law in terms of the microscopic picture of an 'electron gas' inside the conductor material (governed by the laws of (quantum) mechanics).
- *Quantum mechanics*: The ideas and laws of quantum mechanics were developed in the early twentieth century in reaction to new experimental findings that could not be accounted for using the physical laws known at the time. Light waves sometimes seemed to behave like particles, and particles (like the electron) sometimes seemed to behave like waves. To come up with new laws that can capture these phenomena, the pioneers of quantum mechanics relied heavily on mathematical analogies (such as analogies between mechanical trajectories and geometrical optics). An important principle guiding their search was the so-called 'correspondence principle': the laws of quantum mechanics should reproduce the classical laws of mechanics, where the latter laws were known to work (hierarchy of laws).

For any topic within the natural sciences, the history of its development will provide a good example of how the laws of nature and our general understanding have come into existence, and will be very instructive for gaining awareness of this concept.

7.3.4 Energy

As a concept, energy may seem to have a different character than the other three, as it represents a concept 'in' the natural sciences rather than a concept 'about' the natural sciences. However, in our view, the concept 'energy' represents a way of looking at phenomena and systems that is typical of the natural sciences. Energy occurs in many different *forms* (thermal energy, chemical energy, potential energy, electrical energy, radiation, etc) and is of central importance in almost any domain of natural science.

Any natural phenomenon or system can be considered from an 'energy perspective', which often provides important (complementary) insights. The energy concept connects strongly to the other three concepts. For example, the principle of *conservation of energy* provides a powerful way to look at systems, as it can often provide solutions without the need to go into the details of the system (a dimensional analysis works in a similar way). It is also a good example of mathematisation: each *form of energy* corresponds to a certain mathematical form, and it is very hard to explain what energy 'is' without reference to these mathematical forms.

The history of the energy concept also provides a revealing look into the workings of the natural sciences and how natural laws are developed (origin and hierarchy of laws). Finally, energy can be considered in terms of its relevance for society: energy storage, transfer, conversion, and loss. This merits having energy as a recurring theme throughout the natural science courses.

The following micro-concepts exemplify the energy perspective in the natural sciences:

- *Energy conservation*: In any (partial) transformation from one form of energy into another, the total energy before the transformation equals the total energy after the transformation. If energy seems to appear or disappear in a transformation, this indicates that part of the energy is transformed into a form that was not accounted for in the analysis (which can lead to discoveries of new forms of energy).
- *Work and heat* are two distinct pathways of energy *transfer* between systems (or between a system and its environment). Work is energy transferred through an external force deforming or displacing the system, while heat is energy that is transferred spontaneously between a hot object and a cold object. The first and second laws of thermodynamics describe universal limitations relating to these forms of transfer (e.g., no system, such as a car engine, can transform 100% of the heat it receives from combustion into work).
- *Energy minimisation*: Natural systems that interact with their environment tend to evolve towards a state of minimum energy (spontaneously giving up any excess energy to the environment). This idea can be used to find the equilibrium state of a system (for example, one can reason that a soap bubble will have a spherical shape, because this minimises its surface energy for a given volume). The principle of (local) energy minimisation has a deeper origin (origin and hierarchy of laws) in the thermodynamic principle of (global) entropy *maximisation*.

7.4 Connection to the Modelling Process

Solving real-world problems, in any discipline and at any scale, will always have an element of modelling. Simply put, the real world is too complex, and approximations and simplifications must be made to make the problem tangible and solvable.

There are many schemes in literature to describe this modelling process, where we mention the seven-step process by Blum and Leiss (2007) and Blum (2015), as just one example. The seven steps consist of constructing, simplifying/structuring, mathematising, working mathematically, interpreting, validating, and exposing. In the simple scheme in figure 7.1, based on the work of Giere (1999), we have reduced this to a four-step process, where we have combined {Constructing, Simplifying, Mathematising} as a single step 'abstraction', and {Working mathematically} we refer to as 'logical argument', while {Validating} is referred to as 'experiment'.

Our four concepts can be linked to the modelling process in various ways. 'Systems thinking' is an important part of both (mathematical) abstraction and experimentation: What is part of the system and what is not? What phenomena to include? What is the dependent/independent variable?

'Mathematisation of phenomena' directly relates to the right side of the modelling diagram (going from mathematical model to conclusions). In fact, what students learn in this concept is to make the direct link between equations and solutions (e.g., a wave equation gives $sin(\omega t - kx)$), which allows for a quick iteration between the abstraction and the mathematical conclusion. More indirectly, the mathematisation concept also plays a role in the interpretation of mathematics in relation to the real world. A mathematical form may suggest that something will behave like another familiar phenomenon (e.g., since light is a *wave*, it will show interference just like water waves).

The modelling process, as a whole, relates to the 'origin and hierarchy of laws' in that it is a simple representation of the scientific process by which laws are discovered and developed. Furthermore, an awareness of the hierarchy of laws, as previously described, can help in selecting the 'right' laws or level to describe a certain real-world phenomenon. Finally, the concept of 'Energy' links well to both abstraction and drawing real-world conclusions. Energy is one of the most abstract concepts (of mathematical nature) in the natural sciences, yet it is also a concept that, for many people, has very real implications (e.g., paying the energy bill).

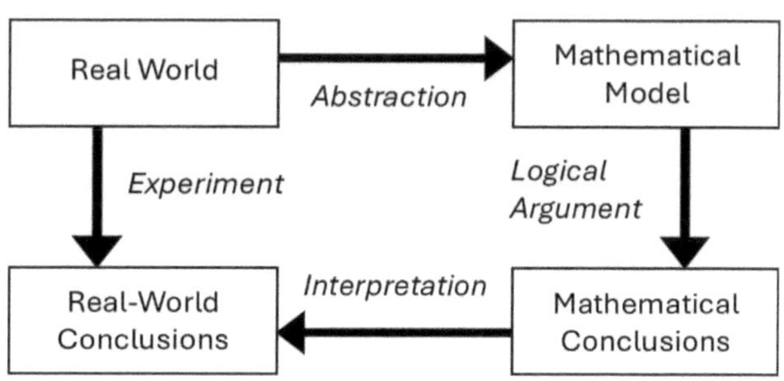

Fig. 7.1 Example of the mathematical modelling process that is used to solve problems in natural science, based on Giere (1999). The concepts connect mainly to the abstraction step and the logical argument step

7.5 Implementation: From Concepts to Courses

Within the context of the bachelor's programme 'Technology, Liberal Arts, and Sciences (ATLAS) at the University of Twente, this approach resulted in an updated natural science curriculum. For the first three semesters, we use a combination of a physical entity (system, particle, field) and a type of phenomenon (transport phenomena, collective phenomena, wave phenomena) through which we can provide different perspectives on the four macro-concepts.

In the fourth semester, the students follow electives outside of the programme, electives that relate to their own academic profile. Together with the students, we analyse these disciplinary courses in terms of the four concepts they have been introduced to. That is, for each course, students will be requested to indicate which laws of nature are used, where they come from, and to what other domains these laws could apply. Also, students will analyse how a systems thinking concept is used in the course. Such a more reflective overview instils a process of zooming out to find a conceptual understanding, which then allows for seeing and using methods and solutions from other fields. It thus helps students in setting up research and designing solutions for challenges. These skills will be immediately applied in the empirical research project of that fourth semester, as well as in the research or design project of the sixth semester (their capstone project).

To illustrate what an implementation of such an approach could look like in a curriculum, we provide more details for the first two semesters of the ATLAS programme.

7.5.1 Semester 1 Natural Science: Systems & Transport Phenomena

While a traditional undergraduate natural science curriculum would often start with a course that recaps and extends on Newtonian Mechanics (statics, kinematics, dynamics), our concepts-based approach suggests a different way to introduce students to the natural sciences, from a broader perspective. Instead of starting with the mechanistic view, we decided to start at a more phenomenological/macroscopic level of describing nature, introducing the empirical works of Fourier (and also Newton) on heat transfer, Ohm on charge transfer, and Poiseuille and Darcy on fluid flow through capillaries or porous media. These topics allow one to cover a wide variety of natural phenomena and concepts that are both close to students' experience and highly relevant in engineering problems and in many scientific fields. In addition, the combination of subjects is especially well-suited to introduce our overarching concepts of 'systems thinking' and 'mathematization of phenomena' (and touch upon the other two concepts as well).

The underlying idea is that natural processes in these different domains (thermal, electrical, and fluid) are described by the same type of equations (the mathematical laws have the same form), which makes it possible to transfer understanding of a situation in one domain to another domain. For example, a water vessel emptying through a porous plug/filter, a cup of hot tea cooling down in a room, and a capacitor discharging through a resistor, all show a similar 'discharging' behaviour characterised by an exponential decay. Other transferable phenomena/concepts that can be readily introduced with these systems are equilibrium and steady state.

The total amount of a quantity (e.g., energy) stored inside some system is indicated as Stock. The Capacity of this system dictates how a change in stock relates to a change in a Potential (for example, temperature). The difference between the system's potential and that of the surrounding drives a Flow of stock between the system and the surrounding from high potential to low potential (quantitatively described by a transport law such as Fourier's law).

To emphasise the similar conceptual/mathematical structure of the three physical domains in the course, we introduced the diagram shown in Fig. 7.2. The terms 'stock' and 'flow' are borrowed from the stock-flow paradigm used in systems engineering and economics (highlighting that the concepts are valuable beyond the natural sciences). A stock is any quantity that can accumulate in a system over time (it is an amount of something 'stored' in the system under study). In the context of economics, this could be people in a country, merchandise in a warehouse, or money in a bank account. In the context of the natural sciences, a stock is a quantity like energy, liquid volume, or charge that represents an amount of something. A net flow of this stock in or out of 'the system' through the system boundaries will result in a change of stock in the system over time (mathematically: rate of change of stock = net flow through boundaries). In short, the relation between stock and flow represents the conservation of amount (relating strongly to the energy concept).

The left-hand side of the diagram (potential -> flow) is where the (domain-specific) dynamics comes in: the transport laws experimentally discovered by Fourier, Ohm, and Darcy/Poiseuille. The most well-known of these laws is probably Ohm's law, which relates a difference in electrical potential (voltage V) to a charge current (flow) I through a resistive wire: $I = V/R$, where R is the resistance of the wire (in turn dependent on the material and geometry of the wire).

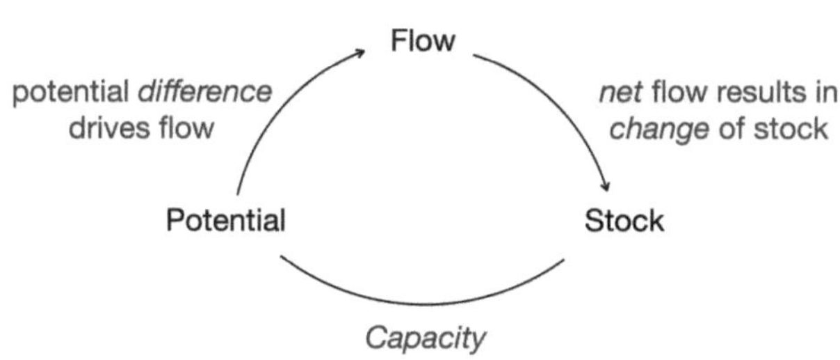

Fig. 7.2 Conceptual description of the transport of thermal energy, mass, or charge between a system and its environment *(The arrows depict causal relations between Potential and Flow and between Flow and Stock. Stock and Potential are correlated (line without arrows) through the concept of Capacity)*

In a precisely analogous way, Fourier's law relates heat flow to a temperature difference, while Darcy/Poiseuille's law relates fluid flow to a pressure difference. Electrical potential, temperature, and pressure are, thus, identified as 'potentials' in these domains, with the common property that these quantities represent a kind of local intensity rather than an amount (e.g., temperature is energy *per* molecule or unit mass). Finally, in many natural systems, there is a direct relation between stock in a system and the potential through which the system interacts with the environment. The concept that relates Stock and Flow is Capacity (the heat capacity of an object exchanging heat, the capacitance of a capacitor in an electrical circuit, or the width of the container that holds a liquid with a certain hydrostatic head).

This conceptual scheme can be readily extended to more complex situations. For example, flow from one system to another (instead of the surrounding), or transport laws that are not linear in the potential difference (such as Torricelli's law). The conceptual model could also be modified for, or contrasted to, situations where potential does not directly drive a flow, but a *change* in flow (relevant for accelerated fluid flows in large pipes, or when coils are involved in an electrical circuit), introducing the additional concept of inertia.

A benefit of the current course setup is that it provides a natural way to connect to introductory math courses on differential equations. All the basic systems in this course are described by basic first-order differential equations, instead of the more advanced second-order differential equations readily encountered in introductory (calculus-based) mechanics. Starting with first-order differential equations makes the transition from high school smoother for students.

The ideas of the course also extend readily to more advanced math concepts. For example, when multiple interacting systems are considered, this naturally leads to systems of first-order equations, which can be approached with the tools often taught in the first year of undergraduate math (such as phase portraits, linear algebra, or even simulation). Extending the ideas even further, to systems that are subdivided into (infinitely) many small subsystems, leads to fields and partial differential equations (and to the methods of numerical approximation of such systems).

7.5.2 Semester 2 Natural Science: Particles & Collective Phenomena

The theme of the second semester is inspired by the fact that in the natural sciences, many phenomena are simply manifestations of the collective behaviour of entities at smaller space and time scales. For instance, pressure is the collective effect of molecules pounding on a wall, and electrical current represents the collective motion of electrons, among others. Another example is the freezing of water. For some reason, the molecules in a liquid 'conspire in concert' to get into an ordered state. The emergence of such larger-scale phenomena out of collective small-scale motion is a prototypical example of the hierarchy of laws and is strongly linked to concept 3.

The course will start with addressing the same phenomena as in the previous semester, but now go into the question of where the governing laws originate from. Beginning with the historical origins and followed by the molecular origins. In particular, the mechanisms for the transport of heat and matter can be explained by kinetic theory. In this, elements of Newtonian Mechanics and Probability Theory are naturally introduced. Numerical simulations will be used as an additional tool for understanding: students will write a small code in which particles move according to Newton's law with simple collision dynamics, so they can 'measure' a collective effect (like the pounding on the wall) and compare with the solutions from the first semester.

From this point on, there are many possible routes, and students pick their own. Students may do similar studies for other 'difference drives flow' phenomena. Students can add complications to the interactions, observe the larger-scale effect, and compare with advanced theories. Students can also choose to study the collective phenomena outside the natural science domain. In the social sciences, biology, and economics, this approach is known as agent-based modelling. Examples of applications are pedestrian dynamics, epidemics, population dynamics, predicting financial markets, and many others, where the interactions of autonomous agents at the microscale are simulated to observe the effect at larger scales.

A second topic that will be addressed this semester is Entropy. The fact that any closed system strives to maximum entropy is basically a manifestation of the statistics of large numbers and applies also to many domains outside physics. After the basics (which will contain elements of Thermodynamics, in which context it was originally introduced), students can again apply the concept to another discipline of their choosing, for instance, economics (Galbraith & Chen, 2025), or dive deeper into the relation between entropy and energy by using ensemble theory, which also connects to systems thinking.

The examples of collective behaviour are so plentiful and present in almost any domain that there is much freedom in the topics to choose from, depending on the project and the interests of students. The mathematics that will be taught alongside the natural science course is statistics and programming.

7.6 Discussion and Conclusion

In this chapter, we have introduced a new setup for the natural science curriculum in an interdisciplinary programme, based on our experience in designing and teaching such a curriculum at the University College Twente over the past 12 years. As a result, our setup is more based on our personal experiences and less on in-depth literature studies into interdisciplinary education research. Instead of using the traditional topics, such as thermodynamics, mechanics, as key elements, we suggest a curriculum that is dictated by concepts of natural science. The motivation for this setup is that we want a focus on creating a 'beta literacy', rather than specific domain knowledge. Our experience is that students find it difficult to apply such knowledge outside the specific context in which it was learned and to move to a conceptual level of understanding. Our expectation is that with our new setup, this would be improved.

Finally, we have introduced macro-concepts as those overarching the natural science domain, where the interdisciplinarity is addressed by making students aware that similar concepts exist in other domains. The question that arises is whether it would be possible and meaningful to base a curriculum on concepts that would be overarching *disciplines*, which are sometimes referred to as super-concepts (Wilson, 2022). Examples listed by Wilson are systems, evolution, hierarchy, and entropy, some of which can be viewed as higher-level generalisations of the macro concepts discussed in this chapter, and could serve as a next step in creating a programme with a fully interdisciplinary mindset.

Acknowledgement We would like to thank all our present and past colleagues who have each inspired, challenged, and shaped our thinking in this: Martin Streng and Ruud van Damme; and Leonie Krab-Hüskens for a critical reading of earlier versions of this manuscript. And most importantly, we want to thank all the students of the 10 cohorts that we had the privilege to teach. Their critical attitude and eagerness to learn have greatly contributed to shaping our ideas about teaching natural sciences in an interdisciplinary programme.

References

Blum, W. (2015). Quality teaching of mathematical modelling: What do we know, what can we do? In *The proceedings of the 12th international congress on mathematical education: Intellectual and attitudinal challenges* (pp. 73–96). Springer.

Blum, W., & Leiss, D. (2007). How do students and teachers deal with modelling problems. In C. Haines, P. Galbraith, & S. Khan (Eds.), *Mathematical modelling: Education, engineering and economics* (pp. 222–231). Woodhead.

Erickson, H. L., & Lanning, L. A. (2013). *Transitioning to concept-based curriculum and instruction: How to bring content and process together*. Corwin Press.

Galbraith, J. K., & Chen, J. (2025). *Entropy economics: The living basis of value and production*. University of Chicago Press.

Giere, R. N. (1999). Using models to represent reality. In *Model-based reasoning in scientific discovery* (pp. 41–57). Springer US.

Goldberg, D. E., Somerville, M., & Whitney, C. (2014). *A whole new engineer. The coming revolution in engineering education*. Threejoy.

Medwell, J., Wray, D., Bailey, L., Biddulph, M., Hagger-Vaughan, L., Mills, G., Oliver, M., & Wake, G. (2019). *Concept-based teaching and learning: Integration and alignment across IB programmes*. University of Nottingham.

Stern, J., Ferraro, K., & Mohnkern, J. (2017). *Tools for teaching conceptual understanding, secondary: Designing lessons and assessments for deep learning*. Corwin Press.

United Nations Development Programme. (n.d.). *Sustainable development goals*. https://www.undp.org/sustainable-development-goals

Wilson, A. (2022). *Being interdisciplinary: Adventures in urban science and beyond*. UCL Press.

Chapter 8
Personal Pursuit: Learning Activities Designed by Students

Fokko Jan Dijksterhuis

8.1 Introduction

The cornerstone of the three-year bachelor's programme Technology, Liberal Arts, and Sciences (ATLAS) is self-directed learning. It puts the student in the lead, with individual academic development of students at the heart of the curriculum. In line with this educational philosophy, students plan their own personal and academic development by specifying their learning needs and activities. A key component of the ATLAS programme is the Personal Pursuit. The Personal Pursuit (PP) is a 6EC (equivalent to 168 hours of work) educational activity in which a student transforms a personal interest into an academic educational project. Students design their own learning by formulating learning outcomes, specifying learning activities, and the final product of the project.

Throughout the existence of the ATLAS programme, the Personal Pursuit has proven to mobilise and stimulate creativity and learning with students in a highly personal and diverse manner. The Personal Pursuit has turned out to be a showpiece of the programme, making an impression with third parties by showing that it is possible to pursue something like a 'personal pursuit' in an academic curriculum. The 2019 report of the first visitation stated:

> The concept of Personal Pursuit (PP) as part of the curriculum steps away from the traditional distinction between curricular and extracurricular activities. PP contributes to enriching the community, as each PP includes an element of outreach to the ATLAS community, and sometimes beyond. (Quality Assurance Netherlands Universities, 2019, p. 24)

F. J. Dijksterhuis (✉)
Knowledge, Transformation & Society (KiTeS), Faculty of Behavioural, Management, and Social Sciences, University of Twente, Enschede, The Netherlands
e-mail: f.j.dijksterhuis@utwente.nl

A. Beldad et al. (eds.), *Designing an Interdisciplinary Engineering Education for Complex Real-World Challenges*, Emergent Directions in Engineering Education 1, https://doi.org/10.1007/978-3-032-20223-9_8

In the Personal Pursuit, students design a learning activity themselves. This chapter discusses how this is organised, what challenges have been addressed, and how the experiences with Personal Pursuit can inspire beyond the original setting. The author of this chapter has been involved in the Personal Pursuit programme since the start of the educational programme and has acted as coordinator for five years.

8.2 Organisation of the Personal Pursuit

The Personal Pursuit is a separate curriculum component in the ATLAS programme, running parallel to the semesters. The Personal Pursuit comprises 6 EC and is obligatory in the first two years of the three-year bachelor's programme, with an optional third one in the final year. In the Personal Pursuit, a student transforms a personal interest into an academic educational project. The Personal Pursuit is organised along the same lines as self-directed learning of the ATLAS programme in general. Students design their own learning by formulating learning outcomes, specifying learning activities, and the final product of the project. In the Personal Pursuit, students take ownership of their learning and academic development. Throughout its existence, the Personal Pursuit has developed into a robust and inspiring form of education.

There are general requirements for the Personal Pursuit. Besides consisting of a feasible project on an academic level, the learning activity should go beyond what is offered in the ATLAS curriculum and other programmes. In other words, the Personal Pursuit cannot be a regular course, training, or internship. The Personal Pursuit should contribute to the student's academic development and the Final Learning Outcomes of the ATLAS programme. In other words, it has to be a part of becoming a new engineer. Examples of personal pursuits include wildlife photography, learning a new language, and producing a documentary.

Finally, the project contains an element of outreach in which the student shares their learning and deliverables with the ATLAS community and beyond. In the proposal for the Personal Pursuit, students specify learning activities, learning outcomes, deliverables, and outreach, which are evaluated by the staff according to the requirements. The evaluation of Personal Pursuit proposals is a deliberative process with a team of teachers, making it hard to give concrete and explicit criteria. Finding a balance between personal interests and academic level is challenging.

8.3 The Personal Pursuit Process

The Personal Pursuit is explicitly grounded in the academic development of the student. This is done through an educational narrative in which the student describes their academic journey and explicates their learning desires. An educational narrative makes it possible to use the educational project as an anchoring point for evaluation and reflection (Wilhelm, 2021). Students express and share the narrative on where they come from in terms of academic development. The educational narrative provides the context for students to articulate their learning needs and interests.

Development of the Personal Pursuit proposal is a collaborative process, emphasising peer learning and interactive feedback. From initial idea to full project proposal, students develop their Personal Pursuit proposals in tutor groups. Tutor groups consist of six students and a tutor from the programme's teaching staff. First- and second-year students are combined, so that the former can draw on the experience of the latter. The tutors organise and guide the peer feedback on ideation and proposal, and provide their own feedback as well. During the period of developing the proposals, the tutor groups meet weekly. All tutors then sit together to evaluate the proposals written by the students. They give written feedback that is discussed by the tutor with the students.

Tutors are free to set up the sessions as they see fit, but a general plan is provided by the Personal Pursuit coordinator. Various steps are involved in writing the personal pursuit proposal.

- Step 1: writing and presenting an educational narrative, learning needs, and topic
- Step 2: formulating learning outcomes, in connection with learning activities and deliverables
- Step 3: formalising learning outcomes; aligning the project with the ATLAS programme
- Step 4: peer and tutor feedback on the proposal

In the process, the tutors oversee that the proposals comply with the requirements of the Personal Pursuit. During the execution of the Personal Pursuit project, tutors act as supervisors, and expert advice can be sought. The tutor groups meet each month or so, with individual supervision on request.

The cycle of planning and evaluation of the Personal Pursuit runs from September until June. At the start of the Fall semester, a workshop is given on ideation and the requirements of the proposal. In September and October, students develop a Personal Pursuit proposal that explicates and justifies the topic, the goals, the activities, and the deliverables. After approval of the proposal, the project is executed from November until June, when it is presented with other Personal Pursuit projects at the ATLAS Expo. After the Expo, the student evaluates their learning with respect to the original goals of the project.

8.4 Educational Setting of the Personal Pursuit

The Personal Pursuit epitomises the self-directed learning (SDL) approach that is fundamental to the educational vision of the ATLAS programme. In SDL, the student is responsible for the entire learning process. The student determines the goals of the learning process and the ways the goals are attained. The ATLAS approach to SDL has been developed in practice, informed by the expertise of the education team of the college and loosely based on educational science literature.

8.4.1 Self-Directed Learning

The concept of SDL originates in the 1970s adult education (Loyens et al., 2008; Jossberger et al., 2010; Saks & Leijen, 2014; Voskamp et al., 2022). Brockett and Hiemstra (1991) identify a distinction between SDL as an instructional process (self-directed learning) and SDL as a personality construct (learner self-direction), introducing 'self-direction in learning' as an umbrella concept (Jossberger et al., 2010, p. 10).

Self-Directed Learning is closely related to, and often confused with, Self-Regulated Learning (SRL). The concept of SRL originates in educational psychology, in particular, the works of Zimmerman and Schunk (1989) and Pintrich (2000). The overall consensus is that SRL is aimed at the micro level and the execution of a learning task; SDL at the macro level, including the setting of a learning task (Jossberger et al., 2010; Saks & Leijen, 2014; Voskamp et al., 2022). SDL and SRL both involve active engagement, goal-oriented behaviour, and student control over learning. In SDL, the learning task is always defined by the student; in SRL, it can be generated by the teacher (Loyens et al., 2008).

The ATLAS implementation of SDL, as described in Chap. 4 of this volume, includes both the teaching process and the personal development aspects. This also applies to the Personal Pursuit: it may be a concrete learning activity, but ownership, autonomy, and responsibility are with the student. The design of learning that the student does in the Personal Pursuit is exemplary for Self-Directed Learning, in particular the definition of learning activities and the formulation of learning outcomes (Brockett, 2006; Voskamp et al., 2022).

8.4.2 Learning Outcomes

In the literature, the concept of 'learning outcome' figures next to terms like 'goal', 'aim', and 'objective'. The concept of 'learning goal' is rather contested. Kennedy (2006) provides an enlightening discussion of the terms aims, objectives (or goals), and outcomes. They argue that 'outcome' is preferred over 'aim', as it employs the perspective of the student and the learning process rather than that of the teacher. The 2015 ECTS Users' Guide (European Union, 2015) also employs the concept of 'learning outcome'. The guide defines learning outcomes as statements of what a learner is expected to know, understand and/or be able to demonstrate after completion of the learning process.

Hussey and Smith (2002) provide a highly critical discussion of the concept of 'learning outcomes', arguing that it is only usable for lectures, meaningless for courses, and wrong for programmes because a level of learning cannot be specified. Taking these considerations into account, in the Personal Pursuit, the concept of 'learning outcome' is used as an anchoring point for (self-)evaluation and reflection.

Personal Pursuit is an individual project to which the debate about 'goals' and 'objectives' may not directly apply, as the debate mainly concerns educational activities predesigned for groups of students. In the literature on learning goals, a distinction is made between goals for assessment and goals for learning (Boor et al., 2021; Chatterjee & Corral, 2017; Kennedy et al., 2006). Bloom-style goals for *assessment* are mainly intended to help teachers organise their education, and goals for *learning* are aimed at communicating with students about teaching plans. In the case of students formulating learning outcomes of their own, the functions of learning and assessment can be considered to be integrated, as both processes are controlled by the student. Letting students think about what evidence can be provided for reaching a learning outcome can help them think about concrete and feasible goals.

Learning outcomes have to be *meaningful* to the students, both concerning the content and concerning the function of the learning goals. Learning outcomes are linked to the topic and ambitions of the project, as well as the overall development and personality of the student. Focusing on the formulation of learning outcomes *per se* can make it into a sterile exercise—a trick that can be mastered, but that is not functional for the project as such or for the student. An important aspect of a learning outcome is that it should enable evaluation and stimulate reflection.

If students do not see the use of proper learning goals, they will not be motivated to think about formulation very deeply. The process of writing a Personal Pursuit proposal is more important than formulating learning outcomes as such. Learning outcomes are a means for learning rather than an end in themselves. The whole trajectory within a student of identifying interests and topics, imagining engaging activities, and awareness of learning is the setting in which learning outcomes can be functional.

8.4.3 Tutoring

The organisation of the Personal Pursuit process emphasises the process rather than instruction. The tutor takes on an SDL role, guiding students in diagnosing learning needs, formulating goals, and planning their learning (Hattie & Timperley, 2007; Jossberger et al., 2010; Timmins, 2008). They are activators rather than facilitators (Hattie, 2012). Teachers in SDL are expected to implement a self-regulatory cycle, in which they assist and empower students to observe their effectiveness (Jossberger et al., 2010; Zimmerman et al., 1996).

The tutors support and encourage students through specific and personalised feedback (Jossberger et al., 2010). The tutor groups are aimed at building and improving the self-directing skills of students by implementing a self-regulatory cycle in the sense of Zimmerman et al. (1996), assisting and empowering students to observe their effectiveness. Implementing feedback in a peer group setting and in interaction between teacher and student ties in with the idea of assessment as learning (Boor et al., 2021; Hattie & Timperley, 2007).

8.5 Conclusion

The process of developing a Personal Pursuit proposal entails a trajectory for a student to identify interests and topics, imagine engaging activities, and raise awareness of learning. In this setting, articulating learning outcomes is functional as reference points for explicating learning activities and deliverables. This is part of a process of developing a proposal that starts with selecting a suitable topic (interest) for learning, identifying activities for learning, and articulating outcomes of learning, which is then fed back into the cycle by specifying the topic, activities, and outcomes. In this iterative process, considering deliverables and evidence is a way to specify the whole project. The process is grounded in an educational narrative, in which a student expresses their interests and ambitions in terms of academic development.

Students designing their own learning can be considered a higher level of self-directed learning. The starting point of Personal Pursuit consists of the needs and desires of the students: what do you want to learn at this moment, and why? The idea of an educational narrative ties in with the contention that the communication with students must be geared towards their motivation: 'What do I want to learn?'. By grounding learning goals in the student's learning journey, they become more meaningful for the student regarding both content and function. It enables the possibility to use the learning goals as anchoring points for evaluation and reflection (Wilhelm, 2021).

8.6 Experiences and Challenges

Since the start of ATLAS, more than 10 years of experience have been built with the Personal Pursuit. This has made it clear that the idea of students designing their own learning is feasible. It has been a vital component of the programme, cherished by both the students and the staff. With the student presentations and the Expo, the Personal Pursuit functions as an outreach beyond the ATLAS community and a showcase of the vision on self-directed learning.

The Personal Pursuit has been developed primarily in practice, drawing on the expertise of the teachers who designed and implemented the ATLAS programme. As a result, the use of educational concepts has been rather loose. For example, the terms 'learning goal' and 'outcome' have been employed rather liberally. The Personal Pursuit proposal template uses the term 'objective', although in the communication with students, the term 'goal' is mostly used. The Personal Pursuit reasons from learning activities, which justifies the use of 'outcome'. The emphasis in the communication to students has been on academic development: think about what you expect to be able to do that you could not do before; how can you show that you can do this?

Students find it difficult to design an educational project. Particularly challenging is formulating proper learning outcomes that ensure educational learning at an academic level. Students find it hard to be specific in formulating outcomes. Often, outcomes are too generic rather than directly relevant to the plan. Students also tend to overestimate the time available for the activities projected. Formulating an outcome that can be evaluated ('measured') is particularly difficult. Students tend to make outcomes vague ('I want to be happy') or indicate an activity rather than an outcome. At the same time, an outcome should not be too strict. One of the main challenges is to make the formulation of learning goals *meaningful* for students, rather than a trick that has to be performed to meet requirements. Over time, the ATLAS staff and the Personal Pursuit team, in particular, have organised the development of Personal Pursuit proposals in a way that addresses these issues and overcomes the basic shortcomings: integrating it within the student's learning journey with a collaborative setting of peer review and tutoring.

A fundamental challenge for the Personal Pursuit is to balance the personal interests of students with the academic level required for the programme. In collaboration with tutors and peers, students are quite capable of considering this aspect, but eventually, the ATLAS staff is responsible for assessing this requirement. The process of approval is deliberative, with various staff members coming to a common evaluation of proposals. Still, the criteria remain general and implicit, making it hard to share them with staff and students. A complicating factor is that teachers have different ideas about what makes a good learning outcome and what the function is (assessment or learning). Nevertheless, the evaluation of Personal Pursuit proposals—like the evaluation in the SDL-cycle in general, as described in chapters 3 and 4 of this volume—is an opportunity to exchange and juxtapose various educational perspectives. The setup of tutor groups and the collective feedback by the tutors has an important role in aligning the ideas of teachers about what a learning outcome is, what a good learning outcome is, and how it functions in the learning process.

Main challenges have existed in the logistics of the Personal Pursuit. Originally, students developed their projects individually and used expert supervisors, both within and outside of the ATLAS programme. It was often difficult to find a suitable supervisor, which could delay the project severely. Moreover, the roles and expectations regarding external supervisors were not always clear, jeopardising the progress of projects. This is part of the reason for the introduction of the tutors, who act as process supervisors. Students are responsible for their own Personal Pursuit project, and some find it hard to maintain the discipline. They procrastinate, and the project is deprioritised because of a full academic schedule. In their academic calendar, four hours are assigned each week to work on their personal pursuit, but this does not always solve the issue of delay.

8.7 Reflection and Outlook

As self-directed learning is increasingly gaining currency in the educational activities and ambitions of the University of Twente and in engineering education in general, the fruits of the project presented here can be of general value. Insights into the challenges of students designing educational projects themselves are valuable for an education that is based on the principles of self-directed learning. The experiences in the ATLAS programme with the Personal Pursuit are valuable for the further improvement and development of the self-directed learning approach.

Students designing their learning can be considered a higher level of SDL. A Personal Pursuit proposal can be considered as a form of educational design, a design by students themselves. The process of writing a Personal Pursuit proposal is a way for students to develop competencies of self-directed learning: learning to learn. The Personal Pursuit, thus functions as a breeding ground for students acquiring SDL competences.

Learning outcomes are an important part of student-designed learning, but not the heart of it, or the end. Formulating learning outcomes functions in self-designed educational projects in several ways. For students, learning outcomes are a means to articulate, consider, and reflect upon what they can learn and how. The development of a proposal is a coherent whole that provides the setting for learning to think about academic development and about 'learning'. For teachers, learning outcomes are a means of supervising students in acquiring competencies in self-directed learning: diagnosing learning needs, devising learning activities, articulating evidence and outcomes, and planning their learning (Hattie & Timperley, 2007; Timmins, 2008). They assist and empower students to self-observe their effectiveness (Zimmerman et al., 1996). Formulating learning outcomes is a tool for fostering self-directed learning.

The setup of the tutor groups made the process of writing a PP proposal efficient and rewarding for both teachers and students. The interaction with and between students and tutors laid the focus on creating and articulating meaningful learning experiences. As a template developed for writing a PP proposal, it can be transferred to other educational contexts at the UT and beyond (3TU, ECIU).

The setup of tutor groups and the collective feedback by the tutors has an important role in aligning the ideas of teachers about what a learning outcome is, what a good learning outcome is, and how it functions in the learning process. It emphasises the importance of teacher collaboration in an SDL setting.

For the further development of the Personal Pursuit as a means to foster self-directed learning, the following aspects can be considered:

- elaborate guidance for the educational narrative;
- develop a rubric that students can use to evaluate each other's goals/proposals (Boor et al., 2021); and
- develop self-assessment of Personal Pursuit (feed forward; assessment as learning, Boor et al., 2021).

As self-directed learning is gaining more and more currency in the educational activities and ambitions of the University of Twente and in engineering education in general, the fruits of the project presented here can be of general value.

References

Boor, I., Gerritsen, D., de Greef, L., & Rodermans, J. (2021). *Meaningful assessment in interdisciplinary education: A practical handbook for university teachers*. Amsterdam University Press.

Brockett, R. G. (2006). Self-directed learning and the paradox of choice. *International Journal of Self-Directed Learning, 3*(2), 27–33.

Brockett, R. G., & Hiemstra, R. (1991). *Self-direction in adult learning: Perspectives on theory, research and practice*. Routledge.

Chatterjee, D., & Corral, J. (2017). How to write well-defined learning objectives. *The Journal of Education in Perioperative Medicine, 19*(4), 1–4.

European Union. (2015). *ECTS users' guide 2015*. https://education.ec.europa.eu/sites/default/files/document-library-docs/ects-users-guide_en.pdf

Hattie, J. (2012). *Visible learning for teachers. Maximizing impact on learning*. Routledge.

Hattie, J., & Timperley, H. (2007). The power of feedback. *Review of Educational Research, 77*(1), 81–112. https://doi.org/10.3102/003465430298487

Hussey, T., & Smith, P. (2002). The trouble with learning outcomes. *Active Learning in Higher Education, 3*(3), 220–233. https://doi.org/10.1177/1469787402003003003

Jossberger, H., Brand-Gruwel, S., Boshuizen, H., & Van de Wiel, M. (2010). The challenge of self-directed and self-regulated learning in vocational education: A theoretical analysis and synthesis of requirements. *Journal of Vocational Education and Training, 62*(4), 415–440. https://doi.org/10.1080/13636820.2010.523479

Kennedy, D., Hyland, Á., & Ryan, N. (2006). *Writing and using learning outcomes: A practical guide (C 3.4-1)*. European Higher Education Area. https://ehea.info/media.ehea.info/file/Qualifications_frameworks/05/0/Kennedy_Writing_and_Using_Learning_Outcomes_597050.pdf

Loyens, S. M., Magda, J., & Rikers, R. M. (2008). Self-directed learning in problem-based learning and its relationships with self-regulated learning. *Educational Psychology Review, 20*(4), 411–427. https://doi.org/10.1007/s10648-008-9082-7

Pintrich, P. R. (2000). The role of goal orientation in self-regulated learning. In M. Boekaerts, P. R. Pintrich, & M. Zeidner (Eds.), *Handbook of self-regulation* (pp. 451–502). Academic.

Quality Assurance Netherlands Universities. (2019). *Report on the bachelor's programme technology and liberal arts and sciences of University College Twente*. QANU.

Saks, K., & Leijen, Ä. (2014). Distinguishing self-directed and self-regulated learning and measuring them in the e-learning context. *Procedia-Social and Behavioral Sciences, 112*, 190–198. https://doi.org/10.1016/j.sbspro.2014.01.1155

Timmins, F. (2008). Take time to facilitate self-directed learning. *Nurse Education in Practice, 8*(5), 302–305. https://doi.org/10.1016/j.nepr.2008.02.004

Voskamp, A., Kuiper, E., & Volman, M. (2022). Teaching practices for self-directed and self-regulated learning: Case studies in Dutch innovative secondary schools. *Educational Studies, 48*(6), 772–789. https://doi.org/10.1080/03055698.2020.1814699

Wilhelm, P. (2021). Fostering quality of reflection in first-year honours students in a bachelor engineering Programme Technology, Liberal Arts & Science (ATLAS). *Journal of Higher Education Theory and Practice, 21*(16), 72–91.

Zimmerman, B. J., & Schunk, D. H. (1989). *Self-regulated learning and academic achievement: Theory, research, and practice*. Springer.

Zimmerman, B. J., Bonner, S., & Kovach, R. (1996). *Developing self-regulated learners. Beyond achievement to self-efficacy*. American Psychological Association.

Chapter 9
Embedding Values, Academic and Professional Competencies in Engineering Education: Learning Lines Approach

Rita Ndagire Kizito

9.1 Introduction

Barnett's (2000) notion of an ongoing transition from "complexity" to "super complexity"—a condition characterised by inherent uncertainty, unpredictability, and perpetual flux—remains highly relevant 25 years later. This persistent evolution means that the conceptual frameworks used to understand and engage with the world are subject to continual contestation. Consequently, preparing the contemporary or "new engineer" necessitates equipping them with competencies vital for this complex future. Nonetheless, such preparation must extend beyond academic and technical knowledge to encompass ethical discernment, personal growth, and well-developed professional competencies.

In a Dutch bachelor's programme, referred to as Technology, Liberal Arts, and Science or ATLAS, we struggled to find effective ways to integrate essential values and attitudes in an engineering curriculum. We wanted to develop students who are academically sound and technically competent, while simultaneously nurturing the personal approaches and attitudes that could transform them into ethically-grounded professional problem solvers.

In an effort to embed these graduate attributes in the curriculum, we encountered some challenges. Initially, we required students to achieve these competencies without allocating any credits to their activities, which led to complaints about additional workload to an already crammed curriculum without any meaningful reward. When we attempted to assign credits, the value-laden activities and those already part of the existing courses did not cohere. The disconnect between what needed to be done, what could be assessed, and how students could engage meaningfully in the learning experiences presented a critical design issue in our approach. While the

R. N. Kizito (✉)
University College Twente, University of Twente, Enschede, The Netherlands

A. Beldad et al. (eds.), *Designing an Interdisciplinary Engineering Education for Complex Real-World Challenges*, Emergent Directions in Engineering Education 1, https://doi.org/10.1007/978-3-032-20223-9_9

idea of having learning lines in the curriculum is conceptually sound, it proved difficult to maintain consistency among the programme's four learning lines (e.g., disciplinary lenses for understanding, processes of creation, communication and collaboration, and academic approach and attitude) with the final learning outcomes while ensuring effective vertical and horizontal integration across the curriculum.

This chapter is an account of the development of the AAA line that aims to establish a cohesive curriculum structure, in which the programme's core values are aligned with the learning lines and final learning outcomes. We wanted to move beyond superficial 'add-ons' of values to create transformative educational experiences in which academic rigour, technical prowess, and values development occurred synergistically rather than combatively. Our educational programme's five core values are self-directed learning, personal development, trailblazing, interdisciplinarity, and community.

In the remainder of the chapter, values-integrated engineering education, the learning lines, and some case studies of successful value integration models are described. This is followed by an introduction to the AAA learning line implementation framework and the syllabus for the learning line. Afterwards, the implementation challenges for the AAA learning line and a few recommendations are discussed.

9.2 The Case for Values-Integrated Engineering Education

Over the past two decades, technological evolution has ranged from subtle shifts to disruptive breakthroughs, including the convergence of information technology, communications, and artificial intelligence (AI). This digital transformation is dissolving traditional disciplinary boundaries (Kamp, 2023), fostering integration between engineering, the exact sciences, and the the social sciences. This fundamental shift requires a critical reimagining of engineering education for the future (Broo et al., 2022).

Concurrently, challenges such as climate change, healthcare accessibility, digital transformation, and sustainable development are demanding professionals who can navigate societal, ethical, and environmental considerations alongside technical problems and who can function beyond the academic and technical domains. From their comprehensive analysis of competency requirements across multiple studies, Passow and Passow (2017) found that effective communication, teamwork, and ethical reasoning emerged as critical professional competencies employers value. Additionally, Souppez (2025) noted that the industry values attributes such as "personal and professional attitude and professional conduct" (p. 138).

Although leading engineering programmes worldwide have moved toward socio-technical integration in their curricula as a response to evolving professional requirements (Graham, 2018), it is not easy to adopt strategies that can alleviate these tensions. One such strategy involves designing an integrated curriculum using the CDIO approach to prepare students to become "well-rounded engineers who understand how to Conceive-Design-Implement-Operate complex, value-added

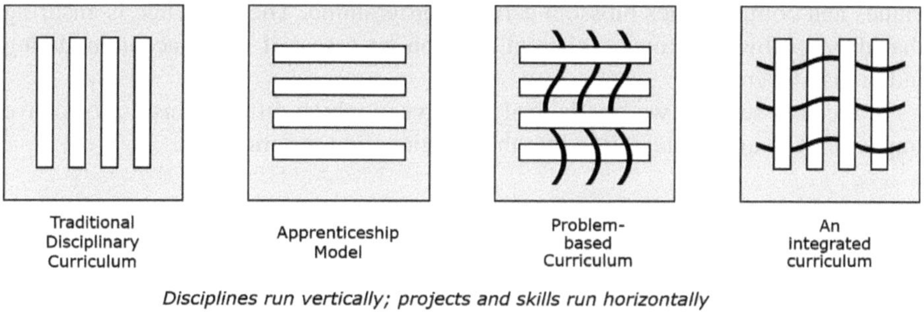

Fig. 9.1 Four approaches to curriculum organisation (adapted from Crawley et al., 2007, p. 98)

engineering products, processes, and systems in a modern, team-based environment" (Crawley et al., 2007, p. 7).

An integrated curriculum organises learning around selected disciplines interwoven with skills, challenges, and projects (Fig. 9.1), contrasting with both traditional disciplinary sequences and pure apprenticeship models. Students learn content through approaches, such as CDIO's design-implement experiences. Problem-based learning is an integrated approach in which students actively engage with problems/projects, integrating required disciplinary content through formal and informal instruction.

These integrated approaches to curriculum design are guided by principles that underpin educational goals, shape learning experiences, and influence the development of student graduate attributes. Values shape these principles. Institutions have different value systems in which values are explicitly defined, implicitly described, or stated as distinct guiding principles in vision and mission statements. Some institutions have well-defined mission statements and educational philosophies.

Regardless of the approach taken, engineering is not value-neutral but is deeply embedded in societal contexts. Engineering decisions have far-reaching consequences—shaping society, the environment, and future generations (Van de Poel & Royakkers, 2011). Therefore, by intentionally integrating values into engineering education, we affirm the inherent social dimension of engineering practice. Learning lines are vehicles that can support the development of value-laden competencies for future engineers.

9.3 Learning Lines as Developmental Pathways for Values Development

The learning lines approach offers a potentially powerful framework for skills and values development when properly conceptualised and implemented. Rather than viewing learning lines as separate curricular components competing with learning outcomes, one can reimagine them as developmental pathways through which values and competencies blossom across a programme. The challenge is ensuring that these pathways remain coherent and can be assessed while accommodating individual differences.

In the next section, we briefly analyse ways in which different institutions have implemented values integration into their engineering programmes.

9.3.1 Case Studies: Successful Values Integration Models

Several institutions have successfully implemented values integration into engineering education, offering valuable lessons for our context. Both Dutch engineering programmes and international examples provide applicable insights.

The University of Twente, in the Netherlands, implemented the Twente Educational Model (TOM), a curriculum structured around thematic modules instead of isolated courses (Visscher-Voerman & Muller, 2017). Each module integrated learning lines, such as 'professional skills' and 'academic approach', with assessment at the module level. This approach fostered integrated development and demonstrated that embedding learning lines within modules is more effective than parallel structures.

Eindhoven University of Technology developed explicit learning trajectories for professional skills development. Meijers and den Brok (2013) outlined their "Engineers for the Future" framework that maps specific learning activities across programme years to ensure progressive development. Particularly noteworthy is the use of anchor courses in each semester, which carried primary responsibility for specific values development, while other courses reinforced and applied these values. This approach balanced dedicated attention with a distributed application.

Delft University of Technology's "Values in Technology" programme explicitly addressed the ethical dimensions of engineering throughout the curriculum. Van de Poel and Royakkers (2011) explained that rather than treating ethics as a separate course or unit, it is better to embed ethical considerations within the technical modules, in which value-sensitive aspects of their disciplines are identified and then incorporated. They preferred values integration, which is discipline-specific rather than generic. Additionally, Aalborg University in Denmark employed problem-based learning to integrate professional values throughout its engineering programmes.

Across the Atlantic, the Olin College of Engineering in the United States pioneered an integrated curriculum centred on design and user-centred approaches. Somerville et al. (2005) presented a "design spine" that serves as the curriculum's core structural framework. This spine features a progressive increase in complexity and student autonomy across each year, seamlessly weaving values, such as ethical consideration and user empathy, into design challenges rather than isolating them. Their approach exemplified how values development could be organically integrated in authentic design activities.

The Rensselaer Polytechnic Institute (RPI), also in the US, implemented a professional development curriculum integrated with technical courses. Their approach included both standalone professional development seminars and embedded activities within the technical courses. Assessment occurred through an electronic portfolio system documenting development across professional competencies. Their model illustrated ways to balance dedicated professional development with technical integration.

A comparison of these approaches revealed several success factors: (a) integration within existing structures rather than parallel systems; (b) progressive development with increasing complexity and authentic application in meaningful contexts, balanced assessment of both technical and professional dimensions; and (c) faculty ownership of the integration process. We used some of these strategies in designing the AAA Line framework.

9.4 Introducing the AAA Learning Line Implementation Framework

The AAA learning line is an important component of the ATLAS programme, which was developed around integrative principles. It employed an innovative approach to engineering education, with the overarching aim of educating 'New Engineers' who can competently analyse socio-technical systems, identify challenges within those systems, and develop appropriate solutions with respect to the context in an interdisciplinary way. This 180-European credit bachelor's programme is founded on the foundational understanding that any technical solution is linked to a social choice and vice versa, reflecting a sophisticated integration of technical competence with social awareness (Beldad & Miedema, 2026).

The programme is organised in six semesters with distinct themes: Semester 1 (Humans and Movement), Semester 2 (Sustainable Cities & Communities), Semester 3 (Engineering for a Resilient World), Semester 4 (Digital Society), Semester 5 (Away & Aware), and Semester 6 (Signature & Celebration). The first four semesters consist of mandatory and elective courses and the semester projects. The fifth semester provides students with the opportunity to pursue educational activities in other universities (either in the Netherlands or abroad) or to do an internship, while students complete their Capstone project (similar to a bachelor thesis) in the sixth semester.

In the first semester, all students took the same foundational courses. As students advanced through the programme, they could choose from a variety of disciplinary courses, often related to their academic interests or disciplinary specialisation. This flexibility allowed them to develop specialised profiles, whether in technical fields (e.g., computer science and engineering, physics) or in broader domains, such as policy development or education.

The engine driving the programme is its assessment strategy, consisting of three phases—the writing of a Personal Development Plan (PDP) at the start of the semester, performance of a Mid-Term Evaluation (MTE), and Self-Evaluation and Reflection (SER), scheduled at the end of a semester. Guiding the educational activities across the six semesters are the five core values that the educational programme envisions as the important attributes its graduates must embody. Table 9.1 briefly describes the five core values.

Although students are required to develop competencies across the four key learning lines—each with its own specific potential for values integration—we will only focus on the AAA learning line in this chapter.

The AAA learning line initially aims to create paths for fostering self-directed learning and personal/professional development responsibly. Through this learning line, students are expected to progressively develop to become ethically aware and autonomous knowledge creators and problem solvers. The programme's assessment mechanism, consisting of the PDP-MTE-SER cycle, as described in the third and fourth chapters of this volume, provided the framework within which the AAA

Table 9.1 UCT-ATLAS core values

Self-directed learning	Through SDL, students are in full control of their learning. This requires an awareness of their strengths and limitations and openness to various opportunities and approaches for learning. Operating in a constantly changing world, students have an open, independent, and critical attitude, and are agile and determined self-directed learners with broad interests.
Personal development	Students are intrinsically motivated to develop themselves holistically and can critically reflect on their own growth. With self-knowledge, drive, and confidence, students develop a sense of their own purpose, and with this inner compass can position themselves in the local and global community.
Trailblazing	Students dare to enter the unknown, pursue radical but appropriate solutions, and are receptive to innovative, creative, and novel ideas. As trailblazers, students are comfortable stepping out of their comfort zones to explore new ways of thinking and doing things. Students are also sufficiently resilient to cope with uncertainties associated with new experiences and unfamiliar challenges.
Interdisciplinarity	Students can competently tackle real-world questions and problems from various disciplinary perspectives and societal contexts. This requires theoretical knowledge and academic and professional skills to understand, identify, and develop appropriate solutions to complex and multidimensional problems. With a strong interdisciplinary mindset, students can effectively collaborate with individuals from a range of disciplines and can translate expert knowledge for different types of stakeholders.
Community	Students are strongly embedded in a community of learners and experts, in their local communities, and in the global community. Students belong to a learning community that values individual contributions, promotes interconnectedness, and respects individual differences. As members of their local communities and of the global community, students are attuned to the challenges confronting those communities and are committed to the collective effort of addressing those challenges.

Learning Line could operate to develop the integrated competencies students are expected to acquire.

9.5 The AAA Learning Line Framework

The AAA learning line is based on a framework that emphasises the development of students' intellectual and character development. This framework also integrates three core pillars: lifewide learning as its educational philosophy (Beard & Jackson, 2011), personal development planning (PDP) as its practical methodology, and developmental progression theory. Lifewide learning emphasises holistic development beyond the classroom, empowering students to create personal growth strategies. PDP provided a structured reflective process for goal-oriented, evidence-based development, typically documented in portfolios. Developmental Progression drew on various theories, such as Chickering and Reisser's (1993) identity formation and the Dreyfus Model of professional development (Honken, 2013), to track students' advancement from novice to competent practitioners.

The framework, shown in Fig. 9.2, resulted from a series of conceptual discussions with a few members of the ATLAS programme's teaching team. Together, we developed a framework from which awareness served as the organising principle for

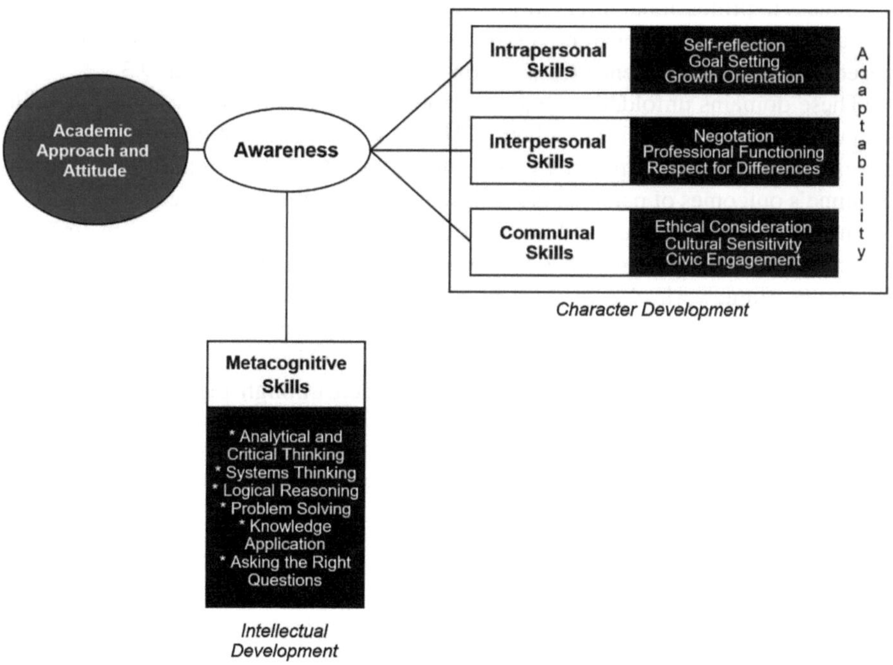

Fig. 9.2 AAA line implementation framework

character and attitudinal development, while knowledge competencies provided the intellectual foundation for professional practice. This ensured that students could develop, not as fragmented specialists, but as integrated professionals capable of ethical reasoning, cultural competence, and systematic inquiry.

Five conceptual foundations emerged as we engaged with the framework, namely:

- *awareness* as an organising principle, which positions self-awareness at the centre, while showing progression to social and cultural awareness;
- *competency integration,* which refers to how thinking skills and character traits develop together;
- *developmental scaffolding,* which emphasises the progression from individual to societal levels of competence;
- *activation through practice,* which underscores the development of competencies through the pursuit of different learning activities; and
- *holistic professional development,* which focuses on the integration of ethical, intellectual, and social dimensions of personal and professional growth.

9.6 Implementing the AAA Line in Practice

The Learning Line was introduced as a structured pathway to embed Academic Approach and Attitude (AAA) across the curriculum—guiding students from self-awareness to professional integration. It was built on three interconnected elements: character development (awareness-based) and intellectual development (knowledge-based), and ethical resilience (integration-focused), as shown in Fig. 9.3.

These domains unfolded across six semesters, with activities designed to build sequentially and spiral in complexity. The AAA syllabus followed a three-stage developmental flow—Explore, Choose/Direct, and Realise—aligning with the programme's outcomes of personal growth, social responsibility, and academic inquiry.

Implementation encouraged collaboration between the AAA learning line and semester coordinators, linking learning activities to semester themes and projects. While the framework effectively connected personal, academic, and professional learning and aligned well with ATLAS values, challenges concerning sustaining integration and consistency across courses are inevitable.

Experimenting with the learning line approach through the AAA framework showed strong outcomes in self-directed learning and personal development, supported by structured goal-setting, reflection, and leadership-focused growth. Trailblazing and community development were moderate, with some innovation and peer engagement, but limited systematic application. Interdisciplinarity remained weak, reflecting minimal cross-disciplinary integration throughout the programme.

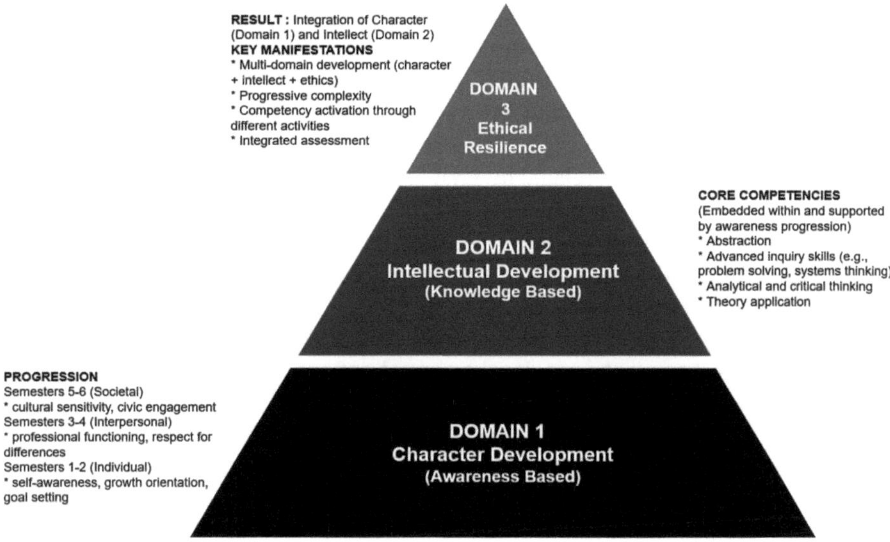

Fig. 9.3 AAA learning line domain

9.7 Implementation Challenges

We analysed the design and implementation experiences, and this analysis revealed that although the AAA Learning Line used a refined framework for personal and professional development, there were significant implementation challenges, which are highlighted in this section.

9.7.1 The Critical Challenges

The workload and credit allocation issue stands as a primary challenge. We initially added value-oriented activities without reducing the original technical content or allocating sufficient credits for the other learning lines. Students understandably viewed these activities as "extra work" without proportional reward. I recall one student commenting: 'We spent hours on these reflections that don't count toward our grade, while we have three technical deadlines the same week'. This tension results from the trade-off between extra workload and the educational benefits of adopting new educational approaches (Harley, 2011). We failed to weigh the initial inputs in the re-design process against the later potential benefits.

Teacher resistance and capacity gaps presented another barrier. University teachers are disciplinary experts but usually lack training in facilitating values development and may feel uncomfortable in these domains. There was some resistance and misinterpretation of what the AAA line development focus entailed, as the roles of the AAA learning line and the PDP-MTE-SER cycle were intertwined. The lack of clear assessment and evaluation criteria and expectations, especially at the last stage, produced some tensions. Czerniewicz and Brown (2009) found similar challenges in their study of four South African universities, identifying the lack of staff training as one of the main barriers to pedagogical innovation.

Coordination challenges related to AAA learning line adoption across the semesters undermined cohesive implementation. The AAA syllabus was well-developed and implemented in the first three semesters, but not in the last three semesters. Dependence on a small team of AAA coordinators rather than the entire team did not help. With the absence of a shared view of how values develop across the curriculum, we created what one student aptly called a 'patchwork of disconnected exercises' rather than a coherent developmental pathway.

The subjective nature of self-awareness and professional identity assessment made it difficult to measure value-laden personal development outcomes and to ensure fair and consistent evaluations across the AAA learning line. The learning activities developed supported the core ATLAS values of self-directed learning and personal development, but failed to truly provide support for trailblazing, interdisciplinarity, and community.

While the syllabus provided a decent progression trajectory from self-awareness to professional competence, the implementation challenges created weaknesses in the realised student journeys along the AAA learning line. Though the framework was pedagogically sound, it required better integration with the context, which, in this case, is the Dutch educational culture and employment market expectations.

The perception of values as less important than technical content remains pervasive. While students acknowledged the importance of professional skills, their behaviour followed assessment priorities. When facing competing deadlines, they directed their efforts toward graded technical assignments rather than the ungraded reflective exercises. The AAA learning line was perceived as an 'add-on' burden to the existing curriculum. Subsequently, student engagement was compliance-based, making it difficult to assess authentic personal development. Student reflection and self-analysis related to the AAA activities were minimal.

The most pervasive challenge was the 'bolt-on' versus 'built-in' tension. When values-driven activities were simply added to existing academic and technical modules without meaningful integration, both students and teachers perceived them as artificial appendages rather than integral components of engineering education. The resulting implementation felt a bit forced and inauthentic, undermining the very values we aimed to develop.

9.8 Lessons Learned and Recommendations

The lessons learned and recommendations reported were related to core design principles, curriculum integration strategies, assessment innovation, student engagement strategies, and teacher development.

9.8.1 Core Design Principles

Employing a mapping tool showing the relationships and variations among the learning lines, programme activity progressions (students' responses to assignments, reflections, and assessment data), and course outcomes enhances the quality of the AAA learning line development. Recent technologies, such as Weever, can be used to create a dashboard that tracks individual student progress and reveal gaps in student learning.

Applying Biggs' (1996) constructive alignment principles cannot be overemphasised. By aligning learning activities, assessment methods, and explicit outcomes, one can ensure that learning is congruent and effective. Subsequently, the academic, technical, and values competencies are aligned and integrated.

In our context, we needed to strengthen the design of the AAA learning line workshops, assignments, and assessments by adding aspects not addressed, such as intercultural competency, research methodologies, and integration of various disciplinary insights in students' Capstone projects.

9.8.2 Curriculum Integration Strategies

Facilitating the implementation of the AAA learning line required building progressive developmental sequences. In our case, we attempted to structure the vertical progression with clear phases starting with foundation-building (explore), followed by introducing values through guided reflection and some application (choose/direct), and transferring values to complex, authentic problems in professional contexts (realise). This progression mapping aligns with seminal contributions of Perry's (1970) stages of intellectual development and Kohlberg's (1984) moral development theory.

In the context of our educational programme, we needed clearer connections with semesters 4 to 6. This could be achieved by:

- identifying complementary topics across concurrent modules and designing single integrative reflections addressing multiple perspectives, reducing redundancy, and supporting coherent curriculum design (Lattuca & Stark, 2009);
- highlighting inherent value moments by identifying points where values naturally emerged from technical challenges (e.g., decisions about balancing cost, safety, ethics) and designing reflective-in-action activities (Schön, 1983) instead of adding separate value activities; and
- embedding values authentically by developing and reflecting in authentic engineering activities, supported by research on moral development (Colby & Sullivan, 2008).

9.8.3 Assessment Innovation

Using assessment activities based on authentic problems or challenges in real-world settings requiring academic and technical expertise and values application (Wiggins, 1998, 2011) was easy to adopt as the ATLAS programme was based on solving challenges/problems.

However, we needed to reinforce meaningful reflection by asking students to describe experiences concretely, examine them through value lenses, and articulate detailed learning implications for deeper metacognitive development. We also needed to strengthen the connection between the AAA learning line and the PDP-MTE-SER cycle.

Strategies that were not fully implemented in our programme included structured peer assessment (Falchikov & Goldfinch, 2000) with clear rubrics to support collaboration and team building. Since students' approaches to learning are fundamentally influenced by assessment practices (Ramsden, 2003), it would be important to allocate credits to values development or values-related outcomes integrated within the academic and technical learning.

Adopting longitudinal portfolio-based assessment using e-portfolio tools, such as PebblePad, is recommended. This corresponds to Barrie's (2007) assertion that assessment should capture development over time rather than at single points.

9.8.4 Student Engagement Strategies

To help students understand why and what values mattered professionally, it is important to articulate clear value propositions using the following features:

- Opportunities for engaging with industry partners who can connect values to career success with concrete examples. A study by Kuh (2009) demonstrated that students engage more deeply when they understand the relevance and purpose of their learning experiences.
- Spaces for student-teacher co-design partnerships allowing both parties to identify and address pain points, leverage unique student perspectives on learning experiences, and generate more engaging approaches. Cook-Sather's (2020) research showed the benefits of student voice in improving engagement and learning outcomes.
- Availability of student choices for personalised learning activities and assessment modalities while balancing consistency with personalisation. Self-determination Theory (Ryan & Deci, 2000) emphasises autonomy as essential for intrinsic motivation and meaningful learning.

Vygotsky's (1978) zone of proximal development concept, adapted for diverse learner needs, is effective in providing flexible scaffolding as it acknowledges that students begin from different starting points and progress through different trajectories at different rates. Tinto's (1997) research on learning communities demonstrated their effectiveness for student engagement and retention. Values development is enriched with student peer groups through project teams and cohort activities.

9.8.5 *Teacher Development*

To effectively embed values-integrated engineering education, comprehensive teacher development should be prioritised. This involves building initial awareness among faculty members through workshops on values frameworks and their roles, followed by establishing cross-disciplinary communities of practice for ongoing collaboration and knowledge sharing. Engaging teachers in co-design for curriculum development ensures authentic integration within their specific contexts. Identifying and supporting champions among early adopters can provide powerful, peer-driven advocacy for sustainable change.

With the advent of AI-supported teaching and learning, it would make sense to invest in developing or accessing digital platforms that can accommodate and integrate personal development plan tracking, resource access, and progress monitoring. Other resource-intensive activities, such as academic advising to strengthen one-on-one coaching relationships throughout the programme, could also be incorporated into this digital platform.

Implementation of values-integrated engineering education requires sustained commitment, a systematic approach, and a willingness to reimagine traditional boundaries between academic, technical, and professional development. These recommendations provide a framework for institutions worldwide to enhance their engineering programmes, while preparing graduates for the complex challenges of contemporary professional practice.

9.9 Conclusion

When properly implemented, the learning lines approach offers a powerful framework for values development, creating structured pathways, while allowing for personalisation and diverse development trajectories. If positioned as an integral component of an engineering curriculum that recognises the inseparable connection between academic, technical competence, and professional/personal development, the AAA learning line has the potential to become a distinguishing feature of any engineering programme.

This can only happen if the implementation challenges are systematically addressed through complete curriculum development, teacher support, and ongoing quality improvement processes. By using this approach, we can develop engineers who are both technically competent and prepared to address complex societal challenges in ethically sound and socially responsible ways.

Transforming engineering education is an ongoing journey. As societal needs, technologies, and professional contexts evolve, our approaches to values integration must likewise adapt. Continuous refinement based on student outcomes, faculty experience, and emerging research will ensure our educational approaches remain relevant and effective.

References

Barnett, R. (2000). Supercomplexity and the curriculum. *Studies in Higher Education, 25*(3), 255–265. https://doi.org/10.1080/713696156

Barrie, S. C. (2007). A conceptual framework for the teaching and learning of generic graduate attributes. *Studies in Higher Education, 32*(4), 439–458. https://doi.org/10.1080/03075070701476100

Beard, C., & Jackson, N. J. (2011). A holistic model of learning and personal development. In N. J. Jackson (Ed.), *Learning for a complex world: A life wide concept of learning, education and personal development* (pp. 39–60). AuthorHouse.

Beldad, A.D., & Miedema, H.A.T. (2026). Introducing a framework for designing an interdisciplinary engineering curriculum: Educating new engineers for complex sociotechnical challenges. *European Journal of Engineering Education, 51*(1), 192–209.

Biggs, J. (1996). Enhancing teaching through constructive alignment. *Higher Education, 32*(3), 347–364. https://doi.org/10.1007/BF00138871

Broo, D. G., Kaynak, O., & Sait, S. M. (2022). Rethinking engineering education at the age of industry 5.0. *Journal of Industrial Information Integration, 25*, 1–8. https://doi.org/10.1016/j.jii.2021.100311

Chickering, A. W., & Reisser, L. (1993). *Education and identity* (2nd ed.). Jossey-Bass Publishers.

Colby, A., & Sullivan, W. M. (2008). Ethics teaching in undergraduate engineering education. *Journal of Engineering Education, 97*(3), 327–338. https://doi.org/10.1002/j.2168-9830.2008.tb00982.x

Cook-Sather, A. (2020). Student engagement through classroom-focused pedagogical partnership: A model and outcomes from the United States. In T. Lowe & Y. El Hakim (Eds.), *A handbook for student engagement in higher education* (pp. 99–109). Routledge.

Crawley, E. F., Malmqvist, J., Östlund, S., & Brodeur, D. R. (2007). *Rethinking engineering education: The CDIO approach* (2nd ed.). Springer.

Czerniewicz, L., & Brown, C. (2009). A virtual wheel of fortune?: Enablers and constraints of ICTs in higher education in South Africa. In S. Marshall, W. Kinuthia, & W. Taylor (Eds.), *Bridging the knowledge divide: Educational technology for development* (pp. 57–76). Emerald Group Publishing.

Falchikov, N., & Goldfinch, J. (2000). Student peer assessment in higher education: A meta-analysis comparing peer and teacher marks. *Review of Educational Research, 70*(3), 287–322. https://doi.org/10.3102/00346543070003287

Graham, R. H. (2018). The global state of the art in engineering education. *Cambridge*, MA: Massachusetts Institute of Technology It can be found here: https://www.rhgraham.org/resources/Global-state-of-the-art-in-engineering-education---March-2018.pdf

Harley, K. (2011). Insights from the health OER inter-institutional project. *Distance Education, 32*(2), 213–227. https://doi.org/10.1080/01587919.2011.584848

Honken, N. (2013, June). Dreyfus five-stage model of adult skills acquisition applied to engineering lifelong learning. In *2013 ASEE Annual Conference & Exposition*

(pp. 23–443). American Society for Engineering Education. https://peer.asee.org/dreyfus-five-stage-model-of-adult-skills-acquisition-applied-to-engineering-lifelong-learning
Kamp, A. (2023). *Navigating the landscape of higher engineering education: Coping with decades of accelerating change ahead*. TU Delft OPEN Publishing.
Kohlberg, L. (1984). *Essays on moral development: Vol. II. The psychology of moral development: The nature and validity of moral stages*. Harper & Row.
Kuh, G. D. (2009). What student affairs professionals need to know about student engagement. *Journal of College Student Development, 50*(6), 683–706.
Lattuca, L. R., & Stark, J. S. (2009). *Shaping the college curriculum: Academic plans in context*. Wiley.
Meijers, A. W. M., & Den Brok, P. J. (2013). *Engineers for the future: An essay on education at TU/e in 2030*. https://pure.tue.nl/ws/portalfiles/portal/3793433/670849533819029.PDF
Passow, H. J., & Passow, C. H. (2017). What competencies should undergraduate engineering programs emphasize? A systematic review. *Journal of Engineering Education, 106*(3), 475–526.
Perry, W. G. (1970). *Forms of intellectual and ethical development in the college years*. Holt, Rinehart and Winston.
Ramsden, P. (2003). The action profile system of movement assessment for self-development. In H. Payne (Ed.), *Dance movement therapy: Theory and practice* (pp. 232–255). Routledge.
Ryan, R. M., & Deci, E. L. (2000). Self-determination theory and the facilitation of intrinsic motivation, social development, and well-being. *American Psychologist, 55*(1), 68–77.
Schön, D. (1983). *The reflective practitioner: How professionals think in action*. Basic Books.
Somerville, M., Anderson, D., Berbeco, H., Bourne, J. R., Crisman, J., Dabby, D., & Zastavker, Y. (2005). The Olin curriculum: Thinking toward the future. *IEEE Transactions on Education, 48*(1), 198–205. https://doi.org/10.1109/TE.2004.842905
Souppez, J. B. R. (2025). Engineering employability skills: Students, academics, and industry professionals' perception. *International Journal of Mechanical Engineering Education, 53*(1), 125–142. https://doi.org/10.1177/03064190231214178
Tinto, V. (1997). Classrooms as communities: Exploring the educational character of student persistence. *The Journal of Higher Education, 68*(6), 599–623. https://doi.org/10.1080/00221546.1997.11779003
Van de Poel, I., & Royakkers, L. M. M. (2011). *Ethics, technology, and engineering: An introduction*. Wiley-Blackwell.
Visscher-Voerman, I., & Muller, A. (2017). Curriculum development in engineering education: Evaluation and results of the Twente Education Model (TOM). In *45th SEFI Annual Conference 2017*.
Vygotsky, L. S. (1978). *Mind in society: The development of higher psychological processes* (Vol. 86). Harvard University Press.
Wiggins, G. (1998). Research news and comment: An exchange of views on "semantics, psychometrics, and assessment reform: A close look at 'authentic' assessments". *Educational Researcher, 27*(6), 20–22. https://doi.org/10.3102/0013189X027006020A
Wiggins, G. (2011). A true test: Toward more authentic and equitable assessment. *Phi Delta Kappan, 92*(7), 81–93. https://doi.org/10.1177/003172171109200721

Chapter 10
Self-Directed Learning (SDL) after Graduation: Graduates' Perspectives on Its Role in Shaping their Attitudes and Behaviours in Personal and Professional Settings

Thiseas Kampouroglou, Ardion Beldad, and Pascal Wilhelm

10.1 Introduction

In a continuously shifting global landscape, university students face conflicting demands from various stakeholders, balancing between institutional requirements, labour market demands, teacher expectations, and their own personal goals for autonomy and independence (Dimitriadis & Kamberlis, 1997; Kromydas, 2017; Morris, 2019; Oddi, 1987). Ease of access to knowledge in the twenty-first century has transformed the ways people learn. The development of artificial intelligence and the existential threat of climate change are indicators of how society is rapidly changing. Hence, academic institutions must prepare their students to adapt to the challenges of their socio-technical environments (World Economic Forum, 2025).

Education can better prepare students for these challenges by moving beyond the traditional one-size-fits-all model and incorporating pedagogies that cultivate students' metacognitive skills (e.g., critical thinking) and their ability to adapt to changes (CEDEFOP, 2017; Morris, 2019). Self-directed learning (SDL), an educational process centred around lifelong learning, could be one such pedagogy. Through an SDL approach to education, learners specify their learning objectives and organise their learning activities in collaboration with experts, and take control of their educational journey to obtain knowledge and acquire skills they desire (Loeng, 2020; Morris, 2019). Research indicates that while implementing SDL in

T. Kampouroglou (✉) · A. Beldad
University College Twente, University of Twente, Enschede, The Netherlands
e-mail: othisseas@protonmail.com

P. Wilhelm
Academy of Applied Psychology and Human Resource Management, Saxion University of Applied Sciences, Deventer, The Netherlands

A. Beldad et al. (eds.), *Designing an Interdisciplinary Engineering Education for Complex Real-World Challenges*, Emergent Directions in Engineering Education 1, https://doi.org/10.1007/978-3-032-20223-9_10

education may be challenging, it can lead to students acquiring enhanced agency, responsibility, and self-regulatory skills.

Morris (2019) views SDL as "a process in which a learner assumes responsibility to control their learning objectives and means in order to meet their personal goals or the perceived demands of their individual context" (p. 634). As an educational philosophy, it postulates that individuals who act based on intrinsic motivation, continuously practice self-reflection, and, with access to the appropriate resources, will be increasingly autonomous, responsible, and self-motivated (Garrison, 1997; Knowles, 1975; Maire et al., 2017). Additionally, SDL aims to help students learn to break the norm of having a fixed path, both in academic and personal contexts, as they become experienced in shaping their future and realising their goals (Yurdal & Toraman, 2023).

Studies into the benefits of SDL have typically relied on qualitative and quantitative data gathered from students currently enrolled in SDL-oriented programmes. However, the long-term effects on students after their SDL education remain largely underexplored. Moreover, while self-directed learning elements are utilised in university curricula globally, few undergraduate university programmes adopt a comprehensive SDL philosophy in their educational design. This results in a notable gap in the literature on the lasting impact of SDL on university graduates. Knowledge of the implications of SDL for the professional and personal lives of graduates matters in understanding the viability of SDL-based education in preparing students for the challenges of the twenty-first century.

The study described in this chapter aims to contribute to the ongoing discussion of the value of SDL. By conducting in-depth interviews with graduates of a three-year bachelor's programme in interdisciplinary engineering—officially referred to as Technology, Liberal Arts, and Sciences or ATLAS—the study aims to understand what an experience with SDL education means for those graduates and how such an experience impacted their attitudes and behaviours in both professional and personal settings. The three research questions below provided the foundation for this study:

(a) What roles does SDL play in the personal and professional lives of graduates of an interdisciplinary engineering bachelor's programme?
(b) What are the perceived effects of SDL on the attitudes and behaviours of graduates of the bachelor's programme?
(c) To what extent did graduates' attitude towards SDL evolve from their time as students to their current status as professionals?

10.2 Literature Review

10.2.1 Conceptualising SDL

While self-directed learning has been heavily researched since its introduction in the 1970s, a generally accepted definition has remained elusive (Kerka, 1994; Loeng, 2020). Knowles (1975) provided the often used definition of SDL, viewing it as a five-step process "in which individuals take the initiative, with or without the help of others, in (1) diagnosing their learning needs, (2) formulating learning goals, (3) identifying human and material resources for learning, (4) choosing and implementing appropriate learning strategies, and (5) evaluating learning outcomes" (p. 18). New researchers to the field face the dilemma of either redefining SDL, thereby adding to the existing "terminological confusion" (Loeng, 2020, p. 3), or basing their research on a fundamentally shared understanding while specifying the particular element under investigation (Candy, 1989; Van der Walt, 2019).

SDL is predicated on a constructivist epistemology, which assumes that reality is not directly understood, but rather perceived through interaction, with knowledge understood as fluid, evolving, and socially constructed (Candy, 1987; Zhang & Zheng, 2013). The central educational implication of constructivism is that learning is a process in which learners build their unique cognitive structures as a result of personal interpretations of their experiences (Candy, 1987; Zhang & Zheng, 2013). Consequently, a constructivist understanding of SDL theory acknowledges that self-directed learning, as a concept, holds different meanings depending on the perspective being considered.

10.2.2 Forms of SDL

In the literature, SDL is viewed either as a process or a personal characteristic. The process conceptualisation aligns with Knowles' (1975) definition, emphasising the learner's autonomy in planning, implementing, and evaluating learning activities, while SDL as a personal characteristic pertains to the underlying traits influencing an individual's tendency to engage with SDL (Lejeune et al., 2023; Raemdonck et al., 2017; Stockdale & Brockett, 2011). Additional distinctions within the process view have been proposed, with Candy (1989) viewing SDL as a personal learning project undertaken outside of formal education and SDL as an instructional approach within educational settings.

10.2.3 Requirements of SDL

To effectively function as a self-directed learner, any student must possess the required attitude and skills. Critical thinking and reflectiveness are indispensable skills for self-directed learners to be able to interrogate their's intrinsic values and assumptions, resulting in the internal change of consciousness integral to SDL (Brookfield, 1985; as cited by Garrison, 1997)

Motivation, and intrinsic motivation in particular, is essential for deep and meaningful learning (Brandt, 2020; Garrison, 1997; Paloş et al., 2025). Garrison (1997) distinguished between two types of motivation: entering motivation, which refers to the motivation required to commit to a particular goal; and task motivation, which refers to the motivation necessary to maintain focus on a learning activity. Entering motivation is influenced by expectancy (learner's self-efficacy, depending on personal characteristics and contextual factors such as institutional barriers) and valence (personal needs and preferences for a learning activity). Although extrinsic motivation may complement intrinsic motivation, it could also limit the learner's willingness to direct learning (Garrison, 1997).

A self-directed learner must also possess a high level of personal responsibility to be able to take ownership of one's learning (Brandt, 2020). In a conventional educational approach, this responsibility lies with teachers who are responsible for designing the education, keeping students focused on the relevant tasks, and evaluating learning outcomes. In SDL education, this responsibility is assumed by the students, who should be motivated to learn, continuously evaluating the extent of their learning (Loeng, 2020; Morris, 2024).

Related to personal responsibility is a sense of autonomy, which refers to the student's ability to take charge of the learning process and "control choices through ongoing reflection and evaluation" (Brandt, 2020, p. 5). It is important to note that autonomous actions happen in a social context and require an awareness of one's environment. Additionally, it involves the ability to independently form a personal identity and make conscious decisions about how to engage with one's surroundings (Brandt, 2020).

Aside from the skills and attitudes mentioned, contextual factors, such as social constraints (e.g., cultural and political climate), teacher characteristics (e.g., personal beliefs, views on sharing authority), and environmental conditions (e.g., institutional structure, classroom setup), are also considered important for the implementation of SDL (Morris, 2024).

10.2.4 Effects of SDL

The positive effects of SDL are well documented in the academic literature. An important point worth noting, however, is that the majority of empirical studies primarily concentrated on the effects experienced by students while still pursuing their education. The implications of SDL for those who experienced it after their education have not been thoroughly understood. What is evident in various studies is the range of benefits students derive from being immersed in SDL.

SDL's effects on students' learning outcomes have received substantial attention. Results of two meta-analyses indicate that SDL processes are positively related to learning outcomes (Aulakh et al., 2025; Sitzmann & Ely, 2011). Specifically, students who utilise SDL methods are reported to achieve high academic success (Edmondson et al., 2012; Sobral, 1997).

Additionally, Rui et al. (2024) found that deep thinking is promoted through SDL's metacognitive effects on learners. Metacognitive skills involve the ability to understand, monitor, and regulate internal learning processes and strategies (Rui et al., 2024). One review found that creativity and curiosity are both bidirectionally related to SDL, with both acting as outcomes of SDL and predictors of students' SDL readiness (Edmondson et al., 2012).

Studies on higher education students have found several metacognitive skills to be bidirectionally associated with SDL (either as outcomes of SDL or motivators of student engagement with SDL). These metacognitive skills involve the ability to understand, monitor, and regulate internal learning processes and strategies (Rui et al., 2024). According to Boyer et al. (2014), SDL projects increased students' confidence and ability, which, the authors further claim, could enhance their self-efficacy. Self-efficacy is an important component of self-regulation and may contribute to SDL's additional outcome of increased perseverance (Sitzmann & Ely, 2011; Sobral, 1997).

Other metacognitive skills, such as self-discipline and emotional control, are also enhanced by SDL (Dahal et al., 2023; Gibbons et al., 1980; Hennecke & Kulkarni, 2024). Additionally, self-directed learners were also found to be adaptable (Dahal et al., 2023), a character trait essential for developing expertise in any subject (Ward et al., 2018). When learners formulate their own educational goals, educational activities increase in perceived value along with the learner's internal locus of control and sense of responsibility, leading to enhanced motivation (Garrison, 1997). Students who have experienced SDL are also reported to be highly inclined to be self-directed throughout their lives (Loeng, 2020; Salleh et al., 2019).

10.3 Method

To address the study's research questions, semi-structured in-depth interviews were conducted with 26 graduates, across nine cohorts (from 2016 to 2024), of an interdisciplinary engineering programme. This method was adopted to obtain diverse insights into the possible long-term effects of self-directed learning.

The interdisciplinary engineering bachelor's programme, officially known as Technology, Liberal Arts, and Sciences or ATLAS, is a three-year liberal arts programme at the University College Twente of the University of Twente, the Netherlands. As previously discussed in Chap. 3 by Wilhelm et al., the programme aims to educate a new breed of engineers who can address complex socio-technical challenges using knowledge and insights from the natural sciences, the social sciences, and mathematics, and can embody the programme's core values of interdisciplinarity, self-directed learning, trailblazing, personal development, and community. Interdisciplinarity is a defining feature of the programme curriculum, with its emphasis on developing students' competencies in the natural sciences (e.g., physics), mathematics, and social sciences (e.g., social psychology, innovation studies), while allowing them to develop their academic interests (similar to disciplinary specialisation).

As Wilhelm et al. described in the third chapter of this volume, self-directed learning is fully embedded in the educational programme's evaluation and assessment through a cycle involving the writing of a personal development plan (PDP) at the start of each semester, and the performance of a midterm evaluation (MTE) for students and the programme to monitor students' academic progress. The final component of this cycle is self-evaluation and reflection (SER), during which students openly discuss with two members of the academic staff how the educational activities specified in their PDP enabled them to meet the semester's learning goals.

In their PDPs, students specify the educational activities (e.g., courses, projects) they intend to pursue to realise the four learning goals of a semester, which are translations of the educational programme's four learning lines: disciplinary lenses for understanding, processes of creation, communication and collaboration, and academic approach and attitude. Through this PDP, students have the autonomy to design their curriculum by making deliberate curricular choices and are encouraged to take responsibility for their learning.

During the early years of the programme, SERs were done in writing. However, this process was considered too one-sided and was eventually transformed into an open discussion between teachers involved in the semester and the student, focusing on their learning experiences, challenges, and plans. An important element of the SER is a reflection on how students' results and experiences in the current semester will impact their plans for the next semester. Figure 10.1 depicts the self-directed cycle adopted by the ATLAS programme across six semesters, as also described in the third and fourth chapters of this volume.

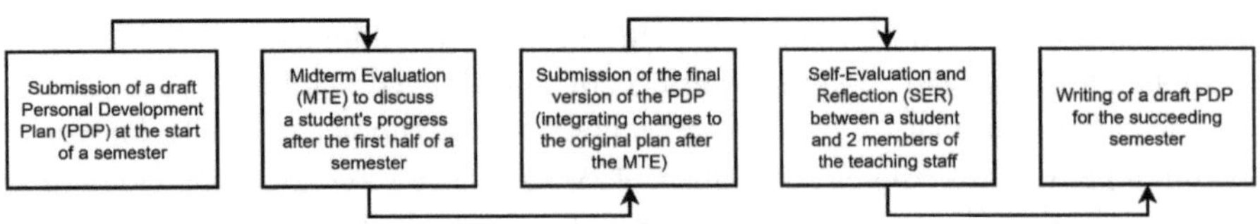

Fig. 10.1 The SDL cycle implemented in the ATLAS programme

10.3.1 Interview Participants

Since its inception in 2013, the ATLAS programme has graduated 217 individuals across nine cohorts, with the most recent batch of students graduating in 2024. To ensure that various insights are collected from each cohort, a convenience stratified sampling method was employed by interviewing at least two graduates per cohort. From bigger cohorts, three to four graduates were invited. Eventually, 26 graduates agreed to participate in the interviews, representing 11.98% of the entire population of graduates.

Invitations to participate in the study were sent through the alumni association of the educational programme, while some were actively approached by email. Those who were already interviewed were also asked to identify fellow graduates who might be interested in participating. Participants' ages ranged from 22 to 31 years old (M = 25.88, SD = 2.69).

Regarding nationality, most participants were Dutch (N = 14), followed by seven participants with a German nationality. A few participants were nationals of Thailand, Indonesia, and Singapore, while another respondent identified as a Spanish national. Of the 26 participants, two have obtained a PhD degree, four are currently working on their PhD, and 11 hold a master's degree. Two respondents reported still working on their master's degree, with the remaining seven respondents having a bachelor's degree as their highest educational qualification. Table 10.1 presents the size of each cohort (number of graduates) and the number of graduates interviewed from each cohort.

10.3.2 Interview Guide

During the interviews, five standard questions were asked to understand how participants experienced SDL during their education, how their attitude towards SDL evolved from their time in the bachelor's programme until the moment they were

Table 10.1 Number of graduates per cohort and the number of participants included in the interviews

Cohort	Graduates	No. of participants interviewed
2016	20	3
2017	21	2
2018	29	4
2019	24	3
2020	35	3
2021	29	2
2022	27	3
2023	24	3
2024	8	3
TOTAL	**217**	**26**

interviewed for this study, and their perception of the role their experience with SDL plays in their professional and personal lives. Presented below are the interview questions.

- How did you experience SDL during your time as a student in ATLAS?
- To what extent did your attitude towards SDL, especially as experienced during ATLAS, change from the moment you graduated from the programme until now?
- To what extent do you think the SDL approach of ATLAS prepared you for your personal and professional life after your bachelor's studies?
- How would you characterise the role of SDL in your life right now?

The semi-structured interview guide enabled the main researcher to pose additional questions based on the interviewees' responses.

10.3.3 Ethical Approval

Before scheduling the interviews, approval from the researchers' university ethics committee was obtained. Before each interview, participants signed a digital consent form describing the study's aims, interview structure, and procedures for data handling. They were also informed of the option to withdraw from the interview at any point and that any data collected from them will be used for research purposes only. Participants consented to the request to audio-record all parts of the conversations.

10.3.4 Data Analysis

Transcription of the interviews was done using Microsoft Teams' software, while coding of all transcripts was performed using ATLAS.ti. Thematic analysis, 'a method for systematically identifying, organising, and offering insights into patterns of meaning (themes) across a data set' (Braun & Clarke, 2006, p. 57), was used, given its appropriateness for understanding the perspectives of research participants and its usefulness in helping researchers handle data in a well-structured manner (Nowell et al., 2017).

Several steps were taken to perform thematic analysis. First, automatically generated transcripts were carefully checked and corrected against the recordings. Meaningful segments of typically one or more sentences in which a participant provided insights into their perspective relevant to the research question were identified, before being assigned a code summarising the segments' main idea. Afterwards, these codes were analysed and inductively grouped into themes of shared experiences. Through iterative cycles of refinement, the codes within each theme were reviewed for relevance to the research questions. Themes were subsequently sorted into broader thematic groups that reflect common topics, which were consolidated into overarching core categories, providing the framework for structuring the results section.

10.4 Results

10.4.1 Understanding of SDL

Participants' understanding of SDL can be positioned on a spectrum, with some viewing it as a subconscious, implicit way of thinking that continuously shapes behaviour and others perceiving it as a consciously applied tool for reflection and planning.

10.4.1.1 SDL as a Way of Thinking

When asked how they view SDL, most participants considered it as something that shapes their mindset and subsequently influences their behaviour. Some explicitly described SDL as a part of their mindset and as *"a subconscious way of thinking" (Participant 8).*

A few viewed SDL as something that has become part of themselves, as evidenced by this remark: *"I don't think it has a super prominent role in the sense that I'm not aware of using it or anything like that, because it's probably become part of me and how I approach things." (Participant 12).*

Related to this way-of-thinking notion is the perception of SDL as a mental framework that helps participants structure their thoughts. This view is captured by the following statement, which subsequently underscores the instrumental nature of SDL as an important tool:

> I mean, if we think of it as a framework to structure thinking, then it could also be a tool, I suppose. (Participant 25)

10.4.1.2 SDL as a Tool

The view of SDL as a tool is also pronounced during the interviews. Some interviewees regarded SDL either as a framework or as a process. As the quotes below emphasise, SDL is like a toolbox that can be easily used in certain cases, such as when they are planning for the future and reflecting on what they want and on ways to achieve it.

> I would compare it to a toolbox, maybe, that I know I have the skills, or at least the experience, and it's something that I know I can always pick up again without much difficulty. (Participant 10)

> It's that sort of continuous conversation [that] I have with myself a lot (and in my relationship), [...] having this sort of knowledge of where are you going and tearing apart the steps to, sort of, get there, that's a skill that I use in my daily life fairly often. (Participant 14)

> I think much of what self-directed learning is to me is self-awareness—what do I want and how do I want to get here. (Participant 26)

Others considered SDL to be a valuable tool for enhancing their sense of personal agency and a feeling of ownership and control over choices and situations. A few equated SDL with having the freedom to decide what and how to learn. The quotes below support the points previously mentioned.

> It's really [about being] able to basically take control of certain situations and [...] determining [for] yourself: How did the past year or X period of time go? What went well? What did not go well? What do I want to do next year of my PhD [...] take more control instead of just going with the flow. (Participant 4)

> .... to really have the freedom and intrinsic motivation to choose what you learn and how you learn. (Participant 23)

In several conversations, SDL was also linked to intrinsic motivation, especially in learning situations, as reflected in the following quote.

> An important part of [SDL] is intrinsic motivation, right?[...] If you're doing this course because you have to because of the requirements of your studies, and you don't really want to do it, then it's really hard to do self-directed learning; then it's not self-directed learning anymore, right? (Participant 18)

The point concerning SDL as a tool for self-development also emerged during the interviews. According to participants, the act of reflection inherent in SDL prompts them to pursue opportunities for self-development. This remark from one participant encapsulates the point stated above.

> In a broad sense [SDL] can also be understood as a continuous development going on and taking an active role [by] identifying where you miss a skill or knowledge and then filling that. (Participant 13)

10.4.2 Effects of SDL

10.4.2.1 Development of Reflection Skills

Participants claimed that their experience with SDL during their three-year bachelor's programme have resulted in several positive outcomes. A dominant theme that emerged during the conversations concerning the effects of SDL is enhanced self-reflection. Most participants admitted that the strong emphasis on SDL in their bachelor's programme contributed to the development of their reflection skills, as exemplified by the remark below.

> I think long term, we probably developed some skills there, some kind of more enhanced reflection, which made having discussions with yourself easier, perhaps. (Participant 16)

According to some participants, these reflection skills made them appreciate their personal growth and accomplishments, as evidenced by this remark.

> I think the reflection part was very useful, because when you're just taking courses, you're looking at the grades that you are receiving. But if you have the reflection part of thinking about how I developed or how this impacts me, it is not just the hard knowledge [I learned], but how it changed what I'm thinking about myself afterwards. (Participant 21)

The value of reflection skills developed through SDL is apparent in specific cases of participants' professional lives, as one participant claimed.

> When writing application letters, putting CVs together, explaining myself, introducing myself, discussing things with people, where at that point you need to take the puzzle pieces and think, OK, how do they fit together? [...] I think in that sense, it's still there but just in a different way. (Participant 14)

10.4.2.2 Increased Sense of Autonomy and Agency

The previous section highlights the view of SDL as a tool enhancing people's sense of autonomy and control over learning processes. In several interviews, it also became clear that SDL has increased participants' sense of autonomy, which manifested in various actions such as goal-setting, decision-making, mindful planning, and maintaining a proactive mindset. Moreover, the sense of autonomy participants acquired from their bachelor's programme has made them *'mature faster' (Participant 11)*.

Goal-setting is noted as an important skill participants acquired from their experience with SDL, with some consequently adopting a goal-oriented mindset. Through SDL, participants recognised the importance of setting realistic and meaningful goals. These points are captured in the following quotes.

> I think having that mindset, where you're thinking about what your goals are and how you're actually spending your time, is really helpful. (Participant 19)

> The choices you have are more like: do I continue with this, or do I move abroad to a different research group where I can be closer to my own target? And then I found that the targets can be more flexible. (Participant 17)

Even after graduating from the programme, the goal-setting skill developed through SDL is still being applied when making professional and personal choices, as one participant quipped.

> I actually thought back to ATLAS and the personal development plan. I was really like, 'OK, think about the next steps, long-term, like the end of my PhD... but then also, what does that mean for this semester? And based on that, I wrote down some major goals, broke them down into smaller goals, and made a plan for myself about what to focus on this semester. (Participant 2)

Alongside the development of goal-setting skills is the sharpening of participants' planning skills, which can be attributed to their experience with writing personal development plans throughout their three-year bachelor's programme. Writing of learning plans as a vehicle for fostering self-directed learning (Du, 2012) is an integral component of the three-year interdisciplinary engineering programme's assessment approach. Hence, it is unsurprising that within a learning environment where deliberate and careful planning is promoted, the programme's graduates will eventually be adept planners, either in professional or personal settings.

> In ATLAS, we are constantly bombarded with … you have to have a plan in place, and it's not just the teachers putting their pressure on you, it's everyone else having a plan, and then you don't …it makes you feel left out, so peer pressure also helps with setting that for you. It's difficult, but it is very, very useful in the end because I see it a lot now in my work. (Participant 8)

The programme's emphasis on students reflecting on how their plans enabled them to reach their semester's learning goals has caused them to be conscious planners, especially as they feel that SDL has empowered them to be in control of their activities.

> I think for me, this empowerment and figuring out the steps I take, how I fill them, and what I do with them, that is in my control and in my agency, and I can sort of take care of that. That's, I think, the most important point I've learned through all the self-reflection, because I actually understand what it does to me. (Participant 6)

Participants also acknowledged how SDL improved their ability to make mindful decisions and to explain the rationale for the decisions made, even if some decisions did not result from a conscious decision-making process. These statements support the points just advanced.

> So, if I have to make a decision, then I can think about several things. I can think about, will I enjoy this decision? Or what will this decision look like to other people? Or what will this decision be? How will I feel about this decision? What do I think I'll have as a choice between A and B? (Participant 25)

> I don't sit down and reflect anymore, but when people like you ask me, Yo, what's up? What did you do? […] that's when I reflect, as I tell people about it, […] sometimes I don't consciously make a decision at the beginning […] but I know why I choose this decision. (Participant 7)

This ability to justify decisions, resulting from years of exposure to SDL, also proved valuable in participants' workplace interactions, especially when the need to explain certain choices arose.

> You can convince your supervisor that you're doing it for a certain reason and that it can actually be interesting. That's something that you're trained to do more or less in ATLAS, [always asking] why are you [choosing] something? And what is the value behind it? (Participant 17)

Furthermore, their experience with SDL during their education has caused participants to be more aware of their actions and maintain a proactive mindset, specifically by learning to '*branch out and take up space*' (Participant 14). The claims below capture the positive outcomes just mentioned.

> Because it worked well for me. I think the whole self-directed learning thing is about being aware of what you're doing and why you're doing it. (Participant 26)

> I think the real mindset is proactive, you know, trying to pick up things, pick up the pace, and get things done. And I think that's very key to ATLAS. (Participant 9)

10.4.2.3 Having a Clear View of Professional Choices

Regular planning and reflection activities, coupled with having the freedom to define their learning activities throughout their bachelor's programme, helped graduates to be future-oriented and to have a clear vision of their professional directions. These are positive outcomes of SDL underscored in the quotes below.

> At some point, you get to realise and gain the maturity to have a bit more of a future vision, and be able to realise where you want to end up and what the necessary steps are to get you there. And I think that the whole self-directed learning cycle, well, I think at least in my experience, it helped me a lot to understand that. (Participant 5)

> In the beginning, you don't have any goals, and I think at the end of ATLAS, I still didn't have a very clear vision of what I wanted to do. So, it was frustrating in that sense, but also by being forced to reflect on it all the time. You do slowly develop a vision or some ideas and some vague sense of direction that becomes clearer and clearer over time, and I think in the last year, actually, I had a pretty good sense of what direction I wanted to go into. (Participant 18)

10.4.2.4 Enhanced Self-Confidence

Across 26 interviews, most participants admitted that their experience with SDL has made them more confident about their abilities and less anxious about trying new things and novel experiences. Additionally, SDL is also seen to have increased their self-efficacy, which cultivated an attitude of being *"like a child that doesn't know what is not possible and suddenly everything is possible"* (P22). These remarks from a couple of participants convey the positive outcomes of SDL.

> In ATLAS, I kind of learned that whatever I want to do I can do if I put my mind to it, and that carried on throughout my master's and also in my job. (Participant 12)

> ... if I didn't have that experience in ATLAS of having so many opportunities to self-direct, I would probably end up the same way as my colleagues who are afraid of doing things, who are afraid of trying something new. (Participant 7)

This heightened sense of self-confidence, some participants claimed, can be attributed to the open nature of their educational programme, which is associated with uncertainty, and its emphasis on 'not fearing to fail'. Constant encounters with uncertainties have apparently cultivated some graduates' confidence in achieving positive results despite the unknowns faced throughout their education.

> Because there are a lot of things that are uncertain and just not fixed, you're forced to deal with this uncertainty and actively engage with the uncertainty [...] but you can probably do it if you work a bit harder, and because you have other skills that help you in that. Maybe in that sense it helped shape this more confident or optimistic view. (Participant 18)

More importantly, their experience with uncertainties, resulting from the open nature of their interdisciplinary engineering education have enhanced their confidence in managing and dealing with the unknown and complex situations, either in their personal or professional lives. This is evident in the remarks below.

> I think I have learned to come to terms with not knowing exactly what I want or how my life in a couple of years will look like [....] overall, (I) kind of make peace with the uncertainty. (Participant 7)

> What I really realised is that ATLAS really gave me a systemic perspective on this world. That there's never just one perspective, there's always multiple. And this always comes with a sort of complexity [...] we can't understand everything completely. Since we can't ever understand everything completely, there's always uncertainty, so uncertainty becomes part of everything in a way. So it becomes a little bit of an acceptable part of living and working. (Participant 6)

10.4.2.5 Well-Developed Critical Thinking Skills

Emphasis on reflection during their years of experiencing SDL has reportedly cultivated participants' critical thinking skills, which remain highly useful in their professional lives, as one participant pointed out.

> I really critically think about basically all my choices both personally and professionally, to see if I'm happy with them, and if not, then prepare for changing direction, and change direction. (Participant 26)

10.4.2.6 Developing New Views on Learning

Undoubtedly, as most participants admitted, SDL has enabled them to develop skills that remain valuable in their professional and personal lives. Throughout the interviews, it also became apparent that their experience with SDL prompted them to have a different view on learning, which manifested through the excitement they felt when undergoing novel learning experiences, as expressed by this participant.

> It was developed there (SDL), so I think that is something that you can apply to a lot of aspects of your life, indeed, personally and professionally. Yeah, just the excitement to take up new things and the ability to teach yourself to learn new things. (Participant 13)

Through SDL, participants also claimed to have taken a more proactive approach to acquiring new knowledge and to identifying opportunities for doing so.

> Having the strategies of how to reach that goal is something I also learned in ATLAS. If I don't know about this topic, or the teacher is not explaining it that well, what alternative paths are there to learning about this topic? (Participant 18)

> To be given the space to explore this in an educational context is a great gift, and it allows you the grace to explore in your own life as well. And I think that it makes it easier for you to seek out opportunities, and it makes you a more compelling person to give opportunities to. (Participant 20)

10.4.3 Issues Associated with SDL

Despite the many positive outcomes participants experienced after years of being immersed in SDL, a few issues have also been raised, suggesting that SDL can be a double-edged sword. One participant likened their development through SDL to becoming *"a spoiled kid who always gets what they want" (Participant 22).*

Interestingly, participants who regularly succeeded in their endeavours, due to the skills acquired through SDL, admitted to having difficulties in handling failures. This is exemplified by the quote below.

> So, if I'm happy, I feel like I've gained it because I worked for it. If I'm not happy, I feel like I can change it, because whatever it takes to change it, I must be able to achieve that by trying to tweak some buttons. [But] if that doesn't work out, I feel like I'm to blame for it and I feel miserable, so it goes both ways. (Participant 26)

The emphasis on constant self-development within the educational programme has also caused some participants to be overly critical of themselves and to overthink situations, as one participant claimed.

> I'm too critical, always. There's always something else I can improve in, and always something more I could do to just reach that goal, and I think that's not so good. I think my education made me an overthinker. (Participant 22)

10.4.4 Attitudes Towards SDL

Across the 26 interviews, it became apparent that the majority of the participants maintain a positive attitude towards SDL, which is attributed to the benefits they have derived from an education predicated on SDL. Presented below are remarks that exemplify the point previously mentioned.

> I have this tool that gave me some more confidence that I can achieve certain things, and I think I'm really grateful for that. (Participant 2)

> I found it very valuable to be forced to think about what you want to learn. (Participant 17)

The positive attitude also results from the autonomy and freedom to explore afforded by SDL, which echoes what has been previously emphasised - that SDL increases participants' sense of autonomy and freedom.

> I enjoyed the idea that I can plan everything myself and I'm given the autonomy and the independence to structure everything in the way I want to, because after all, I am in charge of my own learning, not someone else. (Participant 8)

> [SDL] kind of made me choose more things that felt good, I guess, which was kind of hard, but I really liked it. I had the opportunity to explore. (Participant 1)

A few participants hold a nuanced attitude towards SDL. While they appreciate certain elements of SDL, they are also critical of its implementation in their bachelor's programme, which is viewed to be too 'idealised'. Additionally, the time investment for reflection, despite its perceived benefits, is considered an issue.

> I think [SDL] is how education should be, although it's a bit of an idealised implementation. But I do think that it's something that I think young people should have a chance to experience. (Participant 10)

> It was very time-intensive—all these reflecting things, and I said it might have been a bit too much sometimes, but no, I value it as a good thing, and I applied it in my studies, and I applied it afterwards as well. (Participant 2)

The mismatch between how SDL was implemented in the bachelor's programme and the structure of a more traditional programme has also been raised as a point of criticism.

> I think I understand as to why they would structure it this way, and I think it could have been really nice, but I think in practice it just doesn't really work out because you're also kind of structured within the rest of the university, and everyone's taking exams. Like in a perfect world, maybe, but I don't think it works. (Participant 16)

Participants claimed that their attitude towards SDL from the time they graduated until the moment of the interview had not changed, with some claiming that SDL has become part of their activities.

> I think my perception of it hasn't really changed, other than maybe that I have seen it more in practice. (Participant 13)

Their consistent positive attitude towards SDL is reinforced through comparisons with peers or by their educational and professional experiences. This is captured by the quote below.

> I think if I compare myself to [my] peers in the masters, I've come in with, comparably, high[er] levels of expertise already in the field [showing] that the system works. (Participant 3)

10.5 Discussion, Implications, and Conclusion

10.5.1 Discussion of Results

The results of the focus group discussions somehow validate the dual form of SDL—as a character and as a context-specific process—as described in the literature (Lejeune et al., 2023; Stockdale & Brockett, 2011). These perspectives often overlapped, suggesting that dispositions toward self-direction may foster intentional engagement in SDL processes.

Among the enduring effects of SDL identified by this study, a central outcome was lifelong learning. Supporting theoretical claims that lifelong learning can be fostered through SDL (Boyer et al., 2014; Charokar & Dulloo, 2022), various accounts revealed that it was promoted through the enhanced self-confidence in independent learning, an attitude of ownership over learning, and strategies for individual learning promoted by ATLAS' SDL.

In line with existing evidence, participants reported that their SDL experiences in overcoming challenges improved their reflective skills, self-efficacy, and sense of confidence (Mohiyeddini, 2024; Murad et al., 2010; O'Shea, 2003; Shin et al., 1993). The findings extend this conversation by linking SDL-driven self-confidence to adaptability, comfort with uncertainty, resilience, ability to manage failure, and openness to new experiences (Brandt, 2020). Furthermore, exposure to uncertainty and complexity in education were identified by interviewees as means for developing resilience against uncertainty.

While links among self-esteem, self-efficacy, and SDL have been studied (Hoban & Hoban, 2004), research focusing on SDL's influence on sources of self-worth is lacking. This study identified multiple potential connections, from SDL leading to shifts towards intrinsic sources of self-esteem to enhanced self-confidence leading to the inability to handle failure. An under-researched outcome of SDL may be a connection between success and self-worth, which, according to our results, could be developed through continuous successes and positive feedback, confirming the assertion that overconfidence leads to internal conflict and other adverse effects (Douglas et al., 2023). Future research should, therefore, examine how SDL education prepares (or fails to prepare) its students to cope with failure.

Autonomy, as described by Brandt (2020), is interrelated with personal responsibility, with graduates consistently expressing how they took ownership and made intentional decisions regarding goal-setting and planning, along with personal agency, contributing to studies identifying these benefits in current SDL students (Brandt, 2020; Hewitt-Taylor, 2001; Lejeune et al., 2023; Mohiyeddini, 2024; O'Shea, 2003). These results contribute to the ongoing debate on the connection between SDL and planning, an area where previous research has been inconclusive (Sitzmann & Ely, 2011). Motivation, a primary benefit of SDL (Brandt, 2020; O'Shea, 2003), was also found to persist throughout learners' professional and personal lives.

Echoing Beausaert's (2011) findings concerning the roles of reflectiveness, motivation, and self-directness in the workplace, the skills alumni acquired from their experiences with SDL have prepared them for processes and activities in their professional lives that are comparable to SDL. Additionally, most of the perceived effects of SDL have been identified as the skills most important for success in the twenty-first century (Maire et al., 2017), namely, motivation, self-efficacy, resilience, critical thinking, and creativity.

10.5.2 Implications

Our research findings have implications for educators and institutions intending to implement SDL. Findings reveal that attitude is strongly tied to SDL effectiveness, with the potential to be the strongest factor influencing SDL readiness. Consequently, when applying SDL in an educational institution, both students and the institution itself need to possess a desire for SDL. More specifically, institutions should focus on creating the space for and encouraging SDL. A strong community environment supports effective SDL implementation, lowering the power distance between students and professors, while shifting the importance from individual optimisation towards collective efforts for societal improvement (Brookfield, 1993).

The findings in this paper clearly indicate that SDL can empower learners. Graduates indicated that through SDL, they experienced enhanced motivation, ownership, and agency. These traits are known to foster learning, although the power of SDL might be underestimated.

Second, when lifelong learning becomes the norm, programmes that embrace the SDL approach become the forerunners in preparing students for the world beyond the classroom. Our findings indicate clear and positive transfer effects related to lifelong learning attitudes and behaviours from the educational context to professional life. Educators are, therefore, encouraged to consistently adopt SDL as their educational approach—from the first day of the learner's journey until the endpoint of the trip.

Third, SDL cannot do without appropriate scaffolding. Educators should be aware of the SDL readiness of their incoming students and scaffold appropriately. Related to this need for scaffolding is the introduction of a mentoring approach to ensure that learners are not just left to their own devices but are guided appropriately when challenges associated with SDL arise.

Fourth, for an educational programme to know how SDL is impacting the learning journey of learners, it is important that students' experience with SDL and the possible challenges they encounter when directing their own learning are understood. Through a sensemaking act during SDL, students can obtain clearer insights into the extent of their development—insights that may not be entirely salient to them until they are prompted to reflect on their current experience, rather than doing a purely retrospective reflection. This is when a mentoring approach becomes useful, as this sensemaking activity can be done through informal conversations between a learner and a mentor. Understanding the challenges learners encounter throughout an SDL experience can also inform an educational programme of the appropriate interventions that can be implemented to maximise learners' satisfaction with SDL.

Fifth, programmes implementing SDL should be cautious not to foster a mindset that makes students feel they are fully responsible for their learning and development. This might backfire when circumstances beyond their control would impede the realisation of their plans. Therefore, and in line with the principles of positive education, it is suggested that a focus on well-being is instilled in the programme and that students are made aware that progress should not become a goal in itself.

Self-directed learning is an approach to positive education, characterised by its emphasis not only on developing traditional skills but also on cultivating learners' well-being (Seligman et al., 2009). An education that fosters SDL not only yields knowledge and skills, but also builds character and emphasises individual strengths, such as a sense of autonomy, goal-setting skills, decision-making and planning skills, and the ability to deal with uncertainty. Educators who embrace the idea that education should include both aspects might, therefore, consider adopting SDL as their educational approach. The use of SDL may, therefore, be especially prudent in undergraduate studies, where students are still developing their professional interests. Adoption at the secondary school level might also be considered, given that for learners at that age, self-development (and self-confidence) that allows them to self-regulate and learn effectively is important. It is also a time for learners to decide which direction they want to go in their lives, and SDL could also foster a positive view of learning.

10.5.3 Future Research Directions

The findings reported in this chapter have valuable implications for future research. Considering our study's focus on understanding the role of SDL in the professional and personal lives of alumni, other relevant topics are underexplored. Stemming from our research results are new questions that deserve attention. For instance, how do alumni of an SDL educational programme reconcile potential tensions they experience for being in a professional environment where SDL is not the norm or where individuals are not accustomed to behaviours demonstrating self-directedness? How do alumni handle the downsides associated with having an education rooted in SDL?

10.6 Conclusion

Exploring the long-term influence of undergraduate education based on self-directed learning, this study utilised the lived experiences of students to understand the role of SDL in shaping their attitudes and behaviours in their professional and personal lives. Through semi-structured interviews with alumni from all cohorts of the SDL-oriented educational programme, participants shared their experiences and unique perspectives on how SDL influenced them. The findings contribute empirical examples of the influence of SDL on alumni to the body of literature, identifying avenues for further study and exploring the underrepresented perspectives of SDL-educated alumni.

ATLAS alumni almost unilaterally reported enhanced autonomy, citing SDL in developing skills for acting autonomously. Enhanced goal-setting and planning, decision-making, and the development of clarity of professional desires were reported by most participants, theoretically supported by the dimensions of personal responsibility and autonomy. Graduates' self-development was the primary effect of having an SDL-based education, along with acquired self-directedness throughout their lives. Development of self-confidence, the ability to manage uncertainty, and motivation are among the most prevalent effects of SDL among participants, coinciding with the dimensions of self-regulation and motivation. Moreover, this study supports the predominant understanding of theoretical conceptualisations of SDL, presenting the perspective of alumni in support of the dual conceptualisation of SDL as a process and as a character trait.

References

Aulakh, J., Wahab, H., Richards, C., Bidaisee, S., & Ramdass, P. V. A. K. (2025). Self-directed learning versus traditional didactic learning in undergraduate medical education: A systematic review and meta-analysis. *BMC Medical Education, 25*(1), 1–10. https://doi.org/10.1186/s12909-024-06449-0

Beausaert, S. (2011). *The use of personal development plans in the workplace: Effects, purposes and supporting conditions* [Doctoral dissertation, Maastricht University]. Universitaire Pers Maastricht.

Boyer, S. L., Edmondson, D. R., Artis, A. B., & Fleming, D. (2014). Self-directed learning: A tool for lifelong learning. *Journal of Marketing Education, 36*(1), 20–32. https://doi.org/10.1177/0273475313494010

Brandt, W. C. (2020). *Measuring student success skills: A review of the literature on self-directed learning. 21st century success skills*. National Center for the Improvement of Educational Assessment. https://eric.ed.gov/?id=ED607782

Braun, V., & Clarke, V. (2006). Using thematic analysis in psychology. *Qualitative Research in Psychology, 3*(2), 77–101. https://doi.org/10.1191/1478088706qp063oa

Brookfield, S. (1985). Self-directed learning: A critical review of research. *New Directions for Adult and Continuing Education, 1985*(25), 5–16. https://doi.org/10.1002/ace.36719852503

Brookfield, S. (1993). Self-directed learning, political clarity, and the critical practice of adult education. *Adult Education Quarterly, 43*(4), 227–242. https://doi.org/10.1177/0741713693043004002

Candy, P. C. (1987). *Reframing research into "self-direction" in adult education: A constructivist perspective* [Doctoral dissertation, University of British Columbia]. https://doi.org/10.14288/1.0055826

Candy, P. C. (1989). Constructivism and the study of self-direction in adult learning. *Studies in the Education of Adults, 21*(2), 95–116. https://doi.org/10.1080/02660830.1989.11730524

CEDEFOP. (2017, May 11). *Flexible education and training systems*. https://www.cedefop.europa.eu/en/tools/vet-toolkit-tackling-early-leaving/intervention-approaches/flexible-education-and-training-systems

Charokar, K., & Dulloo, P. (2022). Self-directed learning theory to practice: A footstep towards the path of being a life-long learner. *Journal of Advances in Medical Education & Professionalism, 10*(3), 135–144. https://doi.org/10.30476/JAMP.2022.94833.1609

Dahal, R., Ghimire, B., Karki, D., & Joshi, S. (2023). Elevating job search effectiveness: The significance of self-directed learning and self-control. *Intellectual Economics, 17*, 418–434. https://doi.org/10.13165/IE-23-17-2-08

Dimitriadis, G., & Kamberlis, G. (1997). Shifting terrains: Mapping education within a global landscape. *The ANNALS of the American Academy of Political and Social Science, 551*(1), 137–150. https://doi.org/10.1177/0002716297551001010

Douglas, H. E., Cunningham, M. L., Tisdell, J., & Arneson, J. (2023). The problem with confidence: Too much and too little results in poorer achievement, inner conflict, and social inhibition. *Frontiers in Psychology, 14*, 1–14. https://doi.org/10.3389/fpsyg.2023.960013

Du, F. (2012). Using study plans to develop self-directed learning skills: Implications from a pilot project. *College Student Journal, 46*(1), 223–232.

Edmondson, D. R., Boyer, S. L., & Artis, A. B. (2012). Self-directed learning: A meta-analytic review of adult learning constructs. *International Journal of Education Research, 7*(1), 40–48.
Garrison, D. (1997). Self-directed learning: Toward a comprehensive model. *Adult Education Quarterly, 7*(1), 18–33. https://doi.org/10.1177/074171369704800103
Gibbons, M., Bailey, A., Comeau, P., Schmuck, J., Seymour, S., & Wallace, D. (1980). Toward a theory of self-directed learning: A study of experts without formal training. *Journal of Humanistic Psychology, 20*(2), 41–56. https://doi.org/10.1177/002216788002000205
Hennecke, M., & Kulkarni, P. (2024). Metacognitive knowledge about self-control. *Current Opinion in Psychology, 59*, 1–6. https://doi.org/10.1016/j.copsyc.2024.101861
Hewitt-Taylor, J. (2001). Self-directed learning: Views of teachers and students. *Journal of Advanced Nursing, 36*(4), 496–504. https://doi.org/10.1046/j.1365-2648.2001.02001.x
Hoban, S., & Hoban, G. (2004). Self-esteem, self-efficacy and self-directed learning: Attempting to undo the confusion. *International Journal of Self-Directed Learning, 1*(2), 7–25.
Kerka, S. (1994). *Self-directed learning. Myths and realities*. https://eric.ed.gov/?id=ED365818
Knowles, M. S. (1975). *Self-directed learning: A guide for learners and teachers*. Association Press.
Kromydas, T. (2017). Rethinking higher education and its relationship with social inequalities: Past knowledge, present state and future potential. *Palgrave Communications, 3*(1), 1–12. https://doi.org/10.1057/s41599-017-0001-8
Lejeune, C., Beausaert, S., & Raemdonck, I. (2023). Effective personal development plans contextualized: The role of the autonomy-supportive people manager in sustaining employees' self-directed learning. *Human Resource Development Quarterly, 34*(4), 389–412. https://doi.org/10.1002/hrdq.21492
Loeng, S. (2020). Self-directed learning: A core concept in adult education. *Education Research International, 2020*(1), 1–12. https://doi.org/10.1155/2020/3816132
Maire, Q., Lamb, S., & Doecke, E. (2017). *Key skills for the 21st century: An evidence-based review*. Victoria University.
Mohiyeddini, C. (2024). Self-directed teaching: a holistic framework for educator autonomy. *Frontiers in Medicine, 11*, 1–9. https://doi.org/10.3389/fmed.2024.1479885
Morris, T. H. (2019). Self-directed learning: A fundamental competence in a rapidly changing world. *International Review of Education, 65*(4), 633–653. https://doi.org/10.1007/s11159-019-09793-2
Morris, T. H. (2024). Four dimensions of self-directed learning: A fundamental meta-competence in a changing world. *Adult Education Quarterly, 74*(3), 236–254. https://doi.org/10.1177/07417136231217453
Murad, M. H., Coto-Yglesias, F., Varkey, P., Prokop, L. J., & Murad, A. L. (2010). The effectiveness of self-directed learning in health professions education: A systematic review. *Medical Education, 44*(11), 1057–1068. https://doi.org/10.1111/j.1365-2923.2010.03750.x
Nowell, L. S., Norris, J. M., White, D. E., & Moules, N. J. (2017). Thematic analysis: Striving to meet the trustworthiness criteria. *International Journal of Qualitative Methods, 16*(1), 1–13. https://doi.org/10.1177/1609406917733847
O'Shea, E. (2003). Self-directed learning in nurse education: A review of the literature. *Journal of Advanced Nursing, 43*(1), 62–70. https://doi.org/10.1046/j.1365-2648.2003.02673.x
Oddi, L. F. (1987). Perspectives on self-directed learning. *Adult Education Quarterly, 38*(1), 21–31. https://doi.org/10.1177/0001848187038001003
Paloș, R., Vîrgă, D., & Dediu, R. (2025). Students' approach to learning and their intrinsic motivation to know – The moderating role of psychological needs. *Current Psychology, 44*(12), 11757–11770. https://doi.org/10.1007/s12144-025-07990-x
Raemdonck, I., Thijssen, J., & de Greef, M. (2017). Self-directedness in work-related learning processes. Theoretical perspectives and development of a measurement instrument. In M. Goller & S. Paloniemi (Eds.), *Agency at work: An agentic perspective on professional learning and development* (pp. 401–423). Springer.
Rui, L., Nasri, N., & Mahmud, N. (2024). The role of self-directed learning in promoting deep learning processes: A systematic literature review. *International Journal of Academic*

Research in Progressive Education and Development, 13, 1–24. https://doi.org/10.12688/f1000research.150612.1

Salleh, U. K. M., Zulnaidi, H., Rahim, S. S. A., Bin Zakaria, A. R., & Hidayat, R. (2019). Roles of self-directed learning and social networking sites in lifelong learning. *International Journal of Instruction, 12*(4), 167–182. https://doi.org/10.29333/iji.2019.12411a

Seligman, M. E., Ernst, R. M., Gillham, J., Reivich, K., & Linkins, M. (2009). Positive education: Positive psychology and classroom interventions. *Oxford Review of Education, 35*(3), 293–311. https://doi.org/10.1080/03054980902934563

Shin, J. H., Haynes, R. B., & Johnston, M. E. (1993). Effect of problem-based, self-directed undergraduate education on life-long learning. *CMAJ: Canadian Medical Association Journal, 148*(6), 969–976.

Sitzmann, T., & Ely, K. (2011). A meta-analysis of self-regulated learning in work-related training and educational attainment: What we know and where we need to go. *Psychological Bulletin, 137*(3), 421–442. https://psycnet.apa.org/doi/10.1037/a0022777

Sobral, D. T. (1997). Improving learning skills: A self-help group approach. *Higher Education, 33*(1), 39–50. https://doi.org/10.1023/A:1002997022123

Stockdale, S. L., & Brockett, R. G. (2011). Development of the PRO-SDLS: A measure of self-direction in learning based on the personal responsibility orientation model. *Adult Education Quarterly, 61*(2), 161–180. https://doi.org/10.1177/0741713610380447

Van der Walt, J. (2019). The term "self-directed learning" - Back to Knowles, or another way to forge ahead? *Journal of Research on Christian Education, 28*, 1–20. https://doi.org/10.1080/10656219.2019.1593265

Ward, P., Gore, J., Hutton, R., Conway, G. E., & Hoffman, R. R. (2018). Adaptive skill as the conditio sine qua non of expertise. *Journal of Applied Research in Memory and Cognition, 7*(1), 35–50. https://doi.org/10.1016/j.jarmac.2018.01.009

World Economic Forum. (2025, January 17). *How education can adapt to prepare learners for tomorrow's demands*. World Economic Forum. https://www.weforum.org/stories/2025/01/future-of-education-and-skills/

Yurdal, M. O., & Toraman, Ç. (2023). Self-directed learning, academic achievement and motivation: A meta-analytical study. *Alberta Journal of Educational Research, 69*(2), 233–253. https://doi.org/10.55016/ojs/ajer.v69i2.75098

Zhang, C., & Zheng, G. (2013). Supporting adult learning: Enablers, barriers, and services. In Proceedings of the 14th Annual ACM SIGITE Conference on Information Technology Education (SIGITE '13) (pp. 151–152). ACM. https://doi.org/10.1145/2512276.2512323

Chapter 11
The Challenges of Implementing an Interdisciplinary Engineering Education and the Strategies to Address Them

Ardion Beldad, Pascal Wilhelm, and Jasper Homminga

11.1 Introduction

Complex challenges facing modern societies require effective and innovative solutions emanating from interdisciplinary and transdisciplinary collaborations, as a disciplinary-siloed approach to solving grand challenges has become inadequate. Undoubtedly, what is urgently needed is a concerted effort from various social actors (e.g., experts and academics, citizens, government agencies, the business sector, and nonprofit organisations) to mitigate the undesirable effects of grand challenges. An important point that reverberates throughout this volume is the increasing demand for a new breed of engineers with relevant knowledge and skills for interdisciplinary problem solving. Such a demand has considerable implications for how an engineering education is organised and delivered.

Implementing an interdisciplinary engineering education anchored on a liberal arts educational philosophy and with a strong emphasis on self-directed learning is an important response to calls for a new breed of engineers. With ample freedom to choose in a liberal arts education (Kovačević et al., 2024), engineering students can effortlessly evolve into interdisciplinary professionals, considering the varied opportunities to immerse themselves in the languages, paradigms, and

A. Beldad (✉)
University College Twente, University of Twente, Enschede, The Netherlands
e-mail: a.d.beldad@utwente.nl

P. Wilhelm
Academy of Applied Psychology and Human Resource Management, Saxion University of Applied Sciences, Deventer, The Netherlands

J. Homminga
University College Twente, University of Twente, Enschede, The Netherlands

Faculty of Science and Technology, University of Twente, Enschede, The Netherlands

A. Beldad et al. (eds.), *Designing an Interdisciplinary Engineering Education for Complex Real-World Challenges*, Emergent Directions in Engineering Education 1, https://doi.org/10.1007/978-3-032-20223-9_11

methodologies of different academic disciplines. Furthermore, the emphasis on the equal importance of academic growth, professional skills acquisition, and character building (King et al., 2007) ensures the education of a complete engineer, who, as De Graaff and Ravesteijn (2001) described, is not only intellectually competent but also socially skilled and ethically responsible.

The societal value of interdisciplinary education is undisputable, as highlighted in the first chapter of this volume. At the same time, liberal arts education, as the second chapter underscores, is highly instrumental in educating engineering professionals who can competently address evolving modern-day challenges. Beyond this volume, the importance of interdisciplinary education, particularly for the engineering discipline, has been staunchly advocated. A steady increase in the number of higher educational institutions worldwide that have implemented interdisciplinary engineering programmes attests to the importance of educating a new breed of engineers. Nonetheless, from a curricular and organisational standpoint, implementing an interdisciplinary engineering programme comes with several challenges. In this chapter, we will dissect the prominent challenges confronting an interdisciplinary engineering programme within a Dutch technical university and describe the measures taken to handle them.

11.2 Curricular and Organisational Challenges Associated with the Implementation of an Interdisciplinary Engineering Education

11.2.1 Resolving the Tension Between Breadth and Depth in an Interdisciplinary Engineering Curriculum

The increasing calls to broaden the engineering curriculum by integrating the social sciences (Froyd & Ohland, 2005; Hynes & Swenson, 2013; Sørensen, 2009) and humanities (e.g., O'Neal, 1990; Ruprecht, 1997) signify that resorting to a liberal arts educational philosophy will benefit efforts of educating a new breed of engineers who can solve complex real-world problems. A liberal arts education's emphasis on both broad and deep learning (Hitt et al., 2023) guarantees that students are not only exposed to a range of disciplinary lenses but are also stimulated to pursue specialised knowledge. This suggests that future engineers educated in the liberal arts tradition are not only equipped with the required engineering competencies but also introduced to disciplinary knowledge that is not traditionally prioritised in engineering curricula. Such a broad education with ample space for specialisation is bound to produce 'complete' or 'whole' engineers who are not only academically qualified and technically proficient but also socially and ethically responsible (De Graaff & Ravesteijn, 2001; Hitt et al., 2023).

However, despite the inarguable benefits of educating interdisciplinary engineers, the tension between breadth and depth in an interdisciplinary education is inevitable. The inclusion of mandatory social science and humanities courses in an engineering curriculum triggers questions about the sufficiency of time dedicated to courses traditionally required for completing an engineering degree (e.g., mathematics, physics). This tension is especially acute when considering the three-year duration to complete a bachelor-level engineering degree in a Dutch university. As Self et al. (2019) pointed out, time constraints confronting an undergraduate interdisciplinary program can negatively impact the depth of students' acquisition of foundational knowledge.

Within the Technology, Liberal Arts, and Sciences (ATLAS) programme, the tension is addressed by offering mandatory courses in the social sciences, the natural sciences, and mathematics in the first three semesters of the programme. The first semester has a predefined structure (with only mandatory courses and a semester project) to help students straight from secondary schools acclimate to the self-directed learning philosophy of the programme. After the first semester, students can already take elective courses either provided by the programme or offered by other faculties within the University of Twente.

The number of European credits (EC)—typically 3 ECs equivalent to 84 hours of work (with 1 EC equivalent to 28 working hours, as adopted throughout the European Higher Education Area)—dedicated to the mandatory courses for each of the previously mentioned disciplines restricts those courses' scope and the depth of discussing selected themes. For instance, a social science course within a semester on sustainability can only introduce a bounded set of theories and literature, as opposed to more specialised courses (e.g., psychology of sustainability, sustainability policy) offered by sustainable science programmes.

Hence, students desiring to specialise in sustainable science, for example, must make deliberate choices when filling their elective spaces, especially when considering the disciplinary knowledge gaps they must fill. The spaces for elective courses provided to students guarantee opportunities for knowledge deepening within a discipline. Additionally, during their exchange semester, students can opt to do more specialised courses in a university abroad or within the Netherlands. However, this engenders a dilemma. Should students maintain the interdisciplinary character of their unique educational curriculum with a mixture of courses from both the exact sciences and the social sciences, or should they eventually develop a discipline-oriented curriculum resulting from the urge to specialise?

11.2.2 Ensuring a Personalised Curriculum with a Strong Interdisciplinary Character

The ATLAS programme's open character signifies that students have the autonomy to design their educational curriculum immediately after the first semester. As the semesters progress, the opportunities for students to do elective courses expand, especially from semester 4 onwards, when there are no longer mandatory courses.

However, students' autonomy to define their educational activities must be exercised according to the standard learning goals of each semester. As Wilhelm et al. discussed in the third chapter of this volume, the semester goals are deliberately broadly formulated to enable students to meet the programme's final learning outcomes through a wide range of learning activities. An important point worth mentioning is the role of the learning goals in capturing the programme's four learning lines: disciplinary lenses for understanding, processes of creation (research and design), communication and collaboration, and academic approach and attitude.

The pronounced focus on self-directed learning throughout the educational programme compels students to make deliberate educational choices based on their learning needs. As a component of the programme's self-directed learning approach, students always write a personal development plan (PDP) at the start of each semester to describe how they intend to achieve the semester's learning goals. It is in this plan that students explain their choices of elective courses. In most cases, their choices could be motivated by their intended specialisation or their interests beyond a specific specialisation.

A pressing question, however, concerns ways to ensure that students do not concentrate on overspecialisation when selecting elective courses to preserve the interdisciplinary character of their curricula. Although students of the programme generally have broad academic interests, it is unavoidable that some will strategically select elective courses that will allow them an uncomplicated entry to a discipline-specific master's programme. The programme's prescription that the selection of elective courses is based on academic interests raises another dilemma: to choose courses that provide the competencies required by a master's programme or to follow electives entirely unrelated to the academic path they plan to take - for the sake of interdisciplinarity.

The structure of having mandatory and elective courses guarantees that students are broadening their knowledge base while also deepening their learning in certain academic areas. The inclusion of mandatory social science courses in an engineering curriculum is already instrumental in instilling interdisciplinarity in students' personalised curricula. The elective spaces are valuable avenues allowing students to either broaden or deepen their disciplinary knowledge.

The challenge of maintaining interdisciplinarity surfaces not only in the selection of elective courses but also in students' Capstone projects. Capstone projects, either research- or design-based, are done during the last semester of the programme. Although students are encouraged to conduct interdisciplinary research, this rarely happens due to the discipline-specific nature of the research or design problems students select. Requiring students to perform interdisciplinary research is also unrealistic, as it frequently involves multiple individuals with different disciplinary backgrounds (Tarrant & Thiele, 2017). The expectation that students must already strengthen their academic profile or specialisation in the last three semesters also prods them to focus on discipline-specific research or design problems under the supervision of an experienced researcher from a monodisciplinary educational programme or research group.

To ensure that students' Capstone projects have a degree of interdisciplinary character, the programme requires a combination of supervisors, with one from another faculty with the required disciplinary or content expertise, and the other from within the ATLAS programme, who guide the students in thinking about the interdisciplinary implications of their research findings. In their research reports, students are expected to thoroughly discuss the relevance of their findings for various stakeholders (e.g., government organisations, business and industry, nonprofit sector) and the applicability of the results in other disciplines.

What is apparent is that even in an interdisciplinary educational programme, the tension between monodisciplinary specialisation and transcending disciplines to solve challenges is not easily resolved. The freedom students have in designing their educational curricula also confounds the problem. However, the autonomous role students play in directing their learning must not be curtailed by strict programme requirements of religiously maintaining the interdisciplinary character of their educational activities. In the end, what matters is that students understand the concepts, principles, and theories of different disciplines, and that they know how knowledge from multiple disciplines can be integrated to address grand challenges of our times.

11.2.3 Ensuring the Realisation of the Educational Programme's Final Learning Outcomes with Students' Personalised Curricula

The flexible character of an interdisciplinary educational programme signifies that students are in control of designing their curriculum according to their interests and learning goals (Vincent & Focht, 2011). A significant portion of the ATLAS programme allows students to design their educational curricula. However, this unique feature triggers another critical challenge: how to ensure that students who have distinct educational curricula would ultimately attain the same Final Learning Outcomes (FLOs)? In traditional, monodisciplinary programmes, most courses are mandatory. This makes it easier to align a structure of course goals organised across learning lines with the FLOs. But what if all students take different courses?

The solution to this challenge is the formulation of semester goals. These goals are aligned with the FLOs but are formulated in such a way that they can be attained with different sets of elective courses. The broad formulation makes it impossible to assess them on their own. When students propose in their Personal Development Plans (PDPs) a set of courses for meeting those goals, then the goals can be assessed. Approval of these mandatory plans means that the programme guarantees that the semester goals can be attained with the courses proposed in the PDPs.

Thus, the PDP approval process is critically important to safeguard the quality of assessment. Unlike more traditional programmes, every exam plan of students in the ATLAS programme needs individual approval. This is what happens in the PDP approval process. And this is how every ATLAS student can reach the same FLOs despite having different personalised curricula.

11.2.4 Strengthening the Application and Integration of Disciplinary Knowledge in Tackling Grand Challenges

Problem-based learning, considered an important vehicle for developing interdisciplinary competencies in engineering education (MacLeod & Van der Veen, 2020; Nielsen et al., 2010; Van den Beemt et al., 2020), prescribes the use of a problem as a starting point in the learning process (De Graaf & Kolmos, 2003). The ATLAS programme is structured in six semesters, with the first four associated with specific complex problems or grand challenges (e.g., well-being, sustainability). At the centre of a semester is a team-based project of 9 ECs. Alongside the semester project are the mandatory and elective courses in the first three semesters.

Having a problem at the centre of the learning process necessitates teachers to direct students to learn important concepts in selected core courses for problem-solving (Allen et al., 2011). The role of teachers as guides corresponds to the imperative that in a PBL setting learners are responsible for their own learning, as they specify their learning needs and identify ways to meet those needs (Barrows, 2002). In the ATLAS programme, teachers are not only learning guides but also designers of educational activities instrumental for the implementation of PBL. For instance, the first semester on the theme of well-being consists of teachers from various disciplines (e.g., systems engineering, social and behavioural science, industrial design, physics, and mathematics) responsible for organising a semester project around a theme and the different mandatory courses.

A critical consideration in designing mandatory courses, however, is that the topics introduced have utility for designing solutions. For example, the social science course in the first semester introduces behavioural theories that explain people's decisions to adopt and use a novel technical solution, since users, as argued by Ellen et al. (1991), are bound to resist new technologies. In a mandatory mathematics course, the focus on mathematically modelling complex problems is pronounced, especially since tackling real-world wicked problems is reported to benefit from applying mathematical models (McMillan & Overall, 2016; Tatham & Houghton, 2011).

Despite significant efforts to strengthen the connection between course content and the semester project, students often struggle to utilise and integrate disciplinary knowledge when designing solutions. One approach taken to foster knowledge application at the course level is to develop learning activities and assignments requiring students to tackle a challenge within their semester project. In the social science course, for instance, a typical assignment requires students to critically assess the roles of psychological factors (e.g., trust, personality traits) in enhancing the acceptability of their proposed technical solutions.

In semesters with spaces for elective courses, students are expected to concisely explain the value of their elective courses for the semester projects. This is based on the PBL prescription that students must actively identify the knowledge required for problem-solving (De Graaf & Kolmos, 2003). The form of PBL implemented in the ATLAS programme is at the cross-course level, which, as Chen et al. (2021) described, requires students to work with problems using knowledge from multidisciplinary courses. Using real-world problems as starting points for designing a semester that includes mandatory courses suggests that the course contents must be aligned with the nature of the problem in the semester project. This has profound implications for the mandatory courses offered, considering the potential tension between disciplinary knowledge that must be in the curriculum and the types of knowledge relevant to efforts of solving complex problems.

11.2.5 Provision of Guidance While Students Navigate an Ocean of Courses for Building a Personalised Curriculum

One of the main challenges for students in an interdisciplinary engineering programme such as ATLAS is to make curricular choices. From semester 2 onwards, students must make elective choices to build their curriculum. Students receive instructions on the requirements for these choices and are stimulated to reflect on their current learning needs. Ultimately, they need to make choices from a cornucopia of courses they can take at their home university (or beyond). Students choose the ATLAS programme because of this opportunity.

However, most of them do not know beforehand what curriculum they want to design for themselves. It is in the choice-making that they discover their profile. However, once a choice is made, it is difficult to change it. Courses can be dropped, but this might cause study delays. Therefore, the stakes are high. To facilitate choice-making, students can turn to academic advisors, who are members of the programme's teaching staff. Since almost all teaching staff have a monodisciplinary background, they know much about certain disciplines but not about others. Hence, the challenge concerns how they can better support students when making curricular choices.

One approach the programme employs is to encourage students to solicit specific advice from different teaching staff, depending on their academic interests. For instance, if students are highly interested in the social sciences, they can turn to social science teachers instead of seeking advice from a lecturer in mathematics. Academic staff members are advised to refrain from providing information on specific courses they have limited knowledge of. Instead, they are expected to support students in reflecting on their broader, current interests. Once these interests have become clearer and students can describe their learning needs, it becomes easier for them to select courses.

Advising students to make curricular choices is a skill that must be honed, and academic staff members might require training on asking the right questions. What is important in the choice-making process is that students feel they make choices themselves instead of merely following their teachers' advice. This will foster their sense of autonomy and responsibility for their learning. Nonetheless, according to Robinson and Persky (2020), the sense of autonomy offered by SDL must also be complemented by the provision of guidance from academic staff with the required experience and expertise.

11.2.6 Building an Interdisciplinary Teaching Staff with Monodisciplinary Academics

As interdisciplinary education is organised around complex topics or issues needing viewpoints from a multiplicity of disciplines (Van Goch & Lutz, 2023), having a teaching staff consisting of academics from across a disciplinary spectrum is imperative (Self et al., 2019; Spelt et al., 2009). The primary emphasis on interdisciplinary engineering education throughout the ATLAS programme certainly meets the recommendation of having an interdisciplinary teaching staff. However, this also poses an important challenge concerning team building.

Since teachers in an interdisciplinary programme have their roots in different disciplines (e.g., mathematics, physics, social science), it is not guaranteed that individuals within a teaching team will automatically merge into a coherent, well-functioning entity. There might be misconceptions and prejudice about their respective academic fields and conflicting field-specific conceptions about teaching and learning. Differences in disciplinary epistemologies, discourses, and teaching styles can seriously hamper interdisciplinary collaboration (Bradbeer, 1999), specifically when organising education. Simply putting different disciplines together does not automatically result in integrated teamwork (Meeuwissen et al., 2020).

The challenge here is how to create common ground among teachers in an interdisciplinary programme and to promote productive collaboration. In a study by Meeuwissen et al. (2021), different reciprocal behaviours of team leaders and team members were identified that contribute to effective interdisciplinary team functioning. These behaviours are aimed at promoting inclusiveness. Interdisciplinary teams in which leaders actively engage in coordinating, explicating, inviting, connecting, and reflecting, and team members, apart from participating actively and speaking up, also actively mimic the leader's inclusiveness behaviours, seem to function effectively. This requires openness, flexibility, mutual respect, and persistence to sustain productive collaboration.

An important element is to reach common ground on the basics and the vision of the programme (e.g., student profile, educational approach) and to ensure that those basics function as touchstones for every educational reform. Here, the leader plays an important role in explicating the basics and promoting reflection on the extent to which any educational development or reform still aligns with them.

Another aspect is to support experimentation. Commonly, teachers in interdisciplinary teams have a monodisciplinary background and have little experience with interdisciplinary education from the outset. They need to learn about this type of education through experimentation. One example of such an experiment was a course in ATLAS that focused on sensors. The course was developed by two teachers: one with a background in physics and the other specialising in psychology. The basic idea was that sensors function according to physical principles and that sensor data can be used to measure psychological phenomena (e.g., stress can be measured with the level of skin conductance). A course was designed around this intersection, requiring teachers to learn from each other to run the course (Boor et al., 2021). This approach exemplifies a quote from a famous physicist, Richard Feynman: "If you want to master something, teach it."

In this case, the art of disciplinary teaching was nurtured through collaboration on the teaching job. We argue that to create an interdisciplinary programme, having a coherent, well-functioning teaching team is imperative.

11.2.7 Guaranteeing Access to Disciplinary Experts to Support Students' Learning

As an interdisciplinary programme that uses problem-based learning, the disciplinary knowledge and skills students need to address the challenge at hand are largely unknown beforehand. The programme, thus, requires more disciplinary experts outside of the programme's core teaching staff, who, alongside the students, form an interdisciplinary community of learners, characterised by frequent interaction, interplay, and collaboration among individual members (Yates & Hord, 2023).

A solution that worked well for the ATLAS programme is to have a core team of academics from a range of disciplines capable of taking a zoomed-out view of their disciplines. These academics assume various educational tasks throughout the programme, such as guiding students' self-directed learning, running interdisciplinary semester projects, and developing and delivering mandatory courses. By being part of an interdisciplinary teaching team, they also serve as examples for interdisciplinary collaboration.

This small team is surrounded by a layer of teachers with specific expertise frequently needed for other educational activities and who are affiliated with other faculties within the University of Twente. These teachers (co)teach in certain courses or projects and are frequently part of the students' Capstone project supervision team. They fully understand and support the programme and its goals. The third layer consists of experts in a certain field (e.g., professionals from an organisation) who have the expertise required for supporting students in tackling a challenge in the project.

Within the programme, this spiderweb of teachers has been built through the teaching staff's personal networks, and not through governance. This experience aligns with what the educational programme has learned from other university colleges and other interdisciplinary programmes, although it seems suboptimal from an organisational point of view. It implies that not only does one need to consistently attract teachers who are receptive to new ways of teaching and open to interdisciplinary teaching in interdisciplinary teams, but they also should have an extensive disciplinary network at our university. These requirements limit the pool of teachers suitable for an interdisciplinary educational programme.

11.2.8 Safeguarding Students' Access to Courses from Other (Disciplinary) Programmes

The three-year bachelor's programme in Technology, Liberal Arts, and Sciences was initially established as an educational experiment involving academics from the different faculties of the University of Twente. The programme's model of having mandatory and elective courses signifies that it is also dependent on the disciplinary expertise of academics from other disciplines. While the programme has a core team of teaching staff (representing different faculties and academic disciplines) responsible for mandatory courses and a few electives, the small size of the team prevents it from offering a wide range of elective courses that students need.

Additionally, as an open interdisciplinary programme has no predetermined lines of disciplinary contents and, thus, no predetermined sets of courses, it needs access to disciplinary courses offered by other educational programmes. The students build their own curriculum using that combination of basics from within the ATLAS programme and elective courses from outside. This model has been used since the programme's inception.

More traditional, disciplinary programmes have curricula consisting of courses almost entirely taught by experts in the field, ensuring that all the parts align in terms of content and timing. In practice, teachers oversee one or two courses in such programmes and are mostly unaware of the intricacies of the remainder of the programme. Due to the design of the curriculum, teachers can be certain that students have the necessary knowledge and skills to start the course and need not state or check this. This can be a hurdle for students from open interdisciplinary programmes who only do a limited set of courses in other educational programmes.

The lack of information on prior knowledge and skills makes it difficult for these students to assess if they can start such courses with confidence. Typically, this requires a conversation with the teacher of the course for students to better understand their knowledge gaps and course expectations. These conversations are beneficial for students' development and for the teachers to 'get a feel for' which students they have in their class. It creates confidence in the student-course combination.

The main challenge has been with the programmes themselves. Students intending to build a disciplinary profile will be expected to follow courses from other programmes. Nonetheless, entrance to those courses is often not guaranteed. Certain faculties and educational programmes have rules that determine the suitability of students to follow courses. Concerns regarding the pre-knowledge and skills of ATLAS students are often cited as reasons for denying their requests for course attendance. There are also worries that students pose a challenge to the pre-planned order of the curriculum and the standard procedures used by educational programmes offering the elective courses.

The hurdle for ATLAS students to do elective courses in some educational programmes has not been fully overcome. Through discussions with those programmes and through personal networks, the issue has been partly mitigated, but never at the institutional level. Not addressing this at an institutional level means that every change of management at other educational programmes requires new agreements to allow ATLAS students to follow elective courses that those programmes offer.

11.3 Conclusion

An interdisciplinary engineering education that empowers students to steer their learning is inarguably innovative and valuable. However, the open nature of such an education, as demonstrated by the experience of a three-year bachelor's programme in Technology, Liberal Arts, and Sciences, poses several conundrums. The issues confronting the educational programme since its inception stem from the flexibility of its curriculum and the organisational requirements of an interdisciplinary education (e.g., composition of teaching staff).

The autonomy students have to design their own educational curricula leads to a serious predicament: whether they must maintain a broad curriculum throughout their education or should concentrate on disciplinary specialisation? Additionally, as a consequence of this autonomy, every student will have a distinct educational curriculum. This eventually raises an important question for the educational programme: how to guarantee that the personalised curriculum of students will enable them to realise the programme's final learning outcomes? The process of choice-making related to this autonomy also raises questions concerning adequate provision of guidance to prevent students from being overwhelmed by the abundance of options.

The centrality of problem-based learning within the program is a catalyst for other curricular predicaments. What disciplinary courses should be offered in relation to the type of complex real-world challenge used in a semester project? Should the choice of topics in courses be driven by the theme of the semester project or by an awareness of the knowledge required by disciplines? Should disciplinary knowledge acquired in courses be fully used when tackling the real-world challenge in the semester project? To what extent should a teaching staff intervene when students struggle with knowledge integration for tackling a real-world challenge? These are just a few nagging questions that an interdisciplinary team of teaching staff must address when developing an integrated semester of courses and a project.

Certainly, an interdisciplinary education requires an interdisciplinary teaching staff, which is also associated with a different set of issues. An important hurdle for an interdisciplinary team collaborating on a project is the considerable difference in the team members' disciplinary epistemology, language, and methodologies. Beliefs regarding the hierarchy of disciplines, either explicit or implicit, can also impact relational dynamics within an interdisciplinary team. These issues have profound implications for an interdisciplinary teaching staff responsible for developing and implementing an interdisciplinary engineering education.

Difficulties on the part of students to do courses in other educational programmes are symptomatic of conflicting views on the value of a broad engineering education and misaligned organisational priorities and expectations. What is required to provide interdisciplinary engineering students with a meaningful educational experience is the commitment of various actors—both from within the educational programme and from other programmes—to a shared responsibility of educating a new breed of engineers adequately equipped with knowledge from across disciplinary spectra and nurtured to become responsible members of society. This signifies that the education of a whole engineer, as coined by Hitt et al. (2023), is not an exclusive responsibility of a liberal arts programme but a shared commitment among educational programmes devoted to training young minds for the complex real-world challenges ahead.

Although the challenges discussed in this closing chapter emerge from one educational programme, they can also confront other educational programmes that foster interdisciplinarity and adopt problem-based learning. What is apparent throughout this chapter is that the challenges of implementing an innovative educational programme are perfect opportunities for a learning community of teachers and students to collaboratively learn and develop measures and solutions.

References

Allen, D. E., Donham, R. S., & Bernhardt, S. A. (2011). Problem-based learning. *New Directions for Teaching and Learning, 2011*(128), 21–29. https://doi.org/10.1002/tl.465

Barrows, H. (2002). Is it truly possible to have such a thing as dPBL? *Distance Education, 23*(1), 119–122. https://doi.org/10.1080/01587910220124026

Boor, I., Gerritsen, D., Greef, L. D., & Rodermans, J. (2021). *Meaningful assessment in interdisciplinary education: A practical handbook for university teachers* (1st ed.). Routledge. https://doi.org/10.5117/9789463729048

Bradbeer, J. (1999). Barriers to interdisciplinarity: Disciplinary discourses and student learning. *Journal of Geography in Higher Education, 23*(3), 381–396. https://doi.org/10.1080/03098269985326

Chen, J., Kolmos, A., & Du, X. (2021). Forms of implementation and challenges of PBL in engineering education: A review of literature. *European Journal of Engineering Education, 46*(1), 90–115. https://doi.org/10.1080/03043797.2020.1718615

De Graaf, E., & Kolmos, A. (2003). Characteristics of problem-based learning. *International Journal of Engineering Education, 19*(5), 657–662.

De Graaff, E., & Ravesteijn, W. (2001). Training complete engineers: Global enterprise and engineering education. *European Journal of Engineering Education, 26*(4), 419–427. https://doi.org/10.1080/03043790110068701

Ellen, P. S., Bearden, W. O., & Sharma, S. (1991). Resistance to technological innovations: An examination of the role of self-efficacy and performance satisfaction. *Journal of the Academy of Marketing Science, 19*(4), 297–307. https://doi.org/10.1007/BF02726504

Froyd, J. E., & Ohland, M. W. (2005). Integrated engineering curricula. *Journal of Engineering Education, 94*(1), 147–164. https://doi.org/10.1002/j.2168-9830.2005.tb00835.x

Hitt, S. J., Banzaert, A., & Pierrakos, O. (2023). Educating the whole engineer by integrating engineering and the liberal arts. In A. Johri (Ed.), *International handbook of engineering education research* (pp. 457–476). Taylor & Francis.

Hussain, N., & Ritchey, T. (2011). Wicked problems. *Industrial Pharmacy, 31*, 4–7.

Hynes, M., & Swenson, J. (2013). The humanistic side of engineering: Considering social science and humanities dimensions of engineering in education and research. *Journal of Pre-College Engineering Education Research, 3*(2), 1–12. https://doi.org/10.7771/2157-9288.1070

King, P. M., Brown, M. K., Lindsay, N. K., & VanHecke, J. R. (2007). Liberal arts student learning outcomes: An integrated approach. *About Campus, 12*(4), 2–9. https://doi.org/10.1002/abc.222

Kovačević, M., Dekker, T. J., & van der Velden, R. (2024). Employability development in undergraduate programmes: How different is liberal arts education? *Teaching in Higher Education, 29*(8), 2184–2204. https://doi.org/10.1080/13562517.2023.2212602

MacLeod, M., & Van der Veen, J. T. (2020). Scaffolding interdisciplinary project-based learning: A case study. *European Journal of Engineering Education, 45*(3), 363–377. https://doi.org/10.1080/03043797.2019.1646210

McMillan, C., & Overall, J. (2016). Wicked problems: Turning strategic management upside down. *Journal of Business Strategy, 37*(1), 34–43. https://doi.org/10.1108/JBS-11-2014-0129

Meeuwissen, S. N., Gijselaers, W. H., Wolfhagen, I. H., & Oude Egbrink, M. G. (2020). How teachers meet in interdisciplinary teams: Hangouts, distribution centers, and melting pots. *Academic Medicine, 95*(8), 1265–1273. https://doi.org/10.1097/ACM.0000000000003115

Meeuwissen, S. N., Gijselaers, W. H., van Oorschot, T. D., Wolfhagen, I. H., & Oude Egbrink, M. G. (2021). Enhancing team learning through leader inclusiveness: A one-year ethnographic case study of an interdisciplinary teacher team. *Teaching and Learning in Medicine, 33*(5), 498–508. https://doi.org/10.1080/10401334.2021.1887738

Nielsen, J. D., Du, X. Y., & Kolmos, A. (2010). Innovative application of a new PBL model to interdisciplinary and intercultural projects. *International Journal of Electrical Engineering Education, 47*(2), 174–188. https://doi.org/10.7227/IJEEE.47.2.7

O'Neal, J. B. (1990). The humanities and their effect on engineering education. *IEEE Communications Magazine, 28*(12), 30–35. https://doi.org/10.1109/35.61718

Robinson, J. D., & Persky, A. M. (2020). Developing self-directed learners. *American Journal of Pharmaceutical Education, 84*(3), 292–296. https://doi.org/10.5688/ajpe847512

Ruprecht, R. (1997). Humanities in engineering education. *European Journal of Engineering Education, 22*(4), 363–375. https://doi.org/10.1080/03043799708923468

Self, J. A., Evans, M., Jun, T., & Southee, D. (2019). Interdisciplinary: Challenges and opportunities for design education. *International Journal of Technology and Design Education, 29*(4), 843–876.

Sørensen, K. H. (2009). The role of social science in engineering. In A. Meijers (Ed.), *Philosophy of technology and engineering sciences* (pp. 93–115). North-Holland. https://doi.org/10.1016/B978-0-444-51667-1.50008-2

Spelt, E. J., Biemans, H. J., Tobi, H., Luning, P. A., & Mulder, M. (2009). Teaching and learning in interdisciplinary higher education: A systematic review. *Educational Psychology Review, 21*(4), 365–378. https://doi.org/10.1007/s10648-009-9113-z

Tarrant, S. P., & Thiele, L. P. (2017). Enhancing and promoting interdisciplinarity in higher education. *Journal of Environmental Studies and Sciences, 7*(2), 355–360. https://doi.org/10.1007/s13412-016-0402-9

Tatham, P., & Houghton, L. (2011). The wicked problem of humanitarian logistics and disaster relief aid. *Journal of Humanitarian Logistics and Supply Chain Management, 1*(1), 15–31. https://doi.org/10.1108/20426741111122394

Van den Beemt, A., MacLeod, M., Van der Veen, J., Van de Ven, A., Van Baalen, S., Klaassen, R., & Boon, M. (2020). Interdisciplinary engineering education: A review of vision, teaching, and support. *Journal of Engineering Education, 109*(3), 508–555. https://doi.org/10.1002/jee.20347

Van Goch, M., & Lutz, C. (2023). Scholarly learning of teacher-scholars engaging in interdisciplinary education. *Journal of Interdisciplinary Studies in Education, 12*(SI), 67–90. https://doi.org/10.32674/jise.v12iS1.5313

Vincent, S., & Focht, W. (2011). Interdisciplinary environmental education: Elements of field identity and curriculum design. *Journal of Environmental Studies and Sciences, 1*(1), 14–35. https://doi.org/10.1007/s13412-011-0007-2

Yates, S., & Hord, C. (2023). An interdisciplinary learning community of education and psychology majors. *Education Sciences, 13*(8), 1–16. https://doi.org/10.3390/educsci13080767

Epilogue

Ardion Beldad

Higher educational institutions around the world are increasingly recognising the societal value of interdisciplinary education. Alongside this trend is an upsurge in research interest in the design of interdisciplinary education, the challenges associated with its implementation, and the benefits students and graduates acquire from an education that transcends disciplinary boundaries. This edited volume documents the experience of organising and implementing an interdisciplinary engineering education in a Dutch technical university, while also describing how interdisciplinary/transdisciplinary programmes and liberal arts education are organised in other contexts (e.g., Change Making Project at the University of Amsterdam in the Netherlands; Thapar School of Liberal Arts & Sciences in India). Intensifying calls for innovative approaches to educating a new breed of engineers prompted the launch of a pioneering education - the three-year bachelor's programme Technology, Liberal Arts, and Sciences or ATLAS at the University of Twente, The Netherlands. This interdisciplinary education affords students the autonomy to design their curriculum according to the type of engineer they envision themselves to be.

The societal value of an interdisciplinary education cannot be downplayed. As Boor et al. argue in the first chapter, interdisciplinary and transdisciplinary education (ITD) benefits society by educating future professionals who are future-ready and can competently address contemporary complex challenges. The authors further claim that ITD provides opportunities for generating innovative ideas, especially as they can organically unfold at the intersection of multiple disciplines. However, realising the potential of interdisciplinary education, the authors argue, requires serious commitment to curricular reform and a shift in institutional priorities and societal attitudes.

A. Beldad
University College Twente, University of Twente, Enschede, The Netherlands
e-mail: a.d.beldad@utwente.nl

A. Beldad et al. (eds.), *Designing an Interdisciplinary Engineering Education for Complex Real-World Challenges*, Emergent Directions in Engineering Education 1, https://doi.org/10.1007/978-3-032-20223-9

This call for curricular reform is similarly echoed in the second chapter by Kumar et al., who vigorously defend the role liberal arts education plays in equipping students with invaluable skills to thrive in a highly evolving work landscaperesulting fromrapid technological advancements. The authors contend that despite criticisms regarding its obsolescence and impracticality, liberal arts education is far from irrelevant, given its role in transforming students into future professionals who can think critically, systematically solve problems, clearly articulate views and ideas, and are highly adaptable to change.

Kumar et al. assert that to meet the changing demands of the workforce and the economy, a liberal arts education must consistently rethink its curriculum. Resistance to curricular changes is bound to result in undesirable outcomes. What is specifically needed is commitment from educators to employ innovative teaching approaches, integrate new forms of technology in the learning environment, and create learning experiences that resemble collaborative work in professional settings. At the same time, opportunities for faculty development and fair incentive mechanisms for academic staff merit the serious attention of top management, as the second chapter underscores.

The interdisciplinary engineering education used as a case throughout this volume is primarily hinged on a liberal arts educational philosophy. Juxtaposing the concept of interdisciplinarity with liberal arts seems unsurprising, especially since interdisciplinarity is the core of a liberal arts education (Rhoten et al., n.d.). The authors further emphasise the important role liberal arts education plays in cultivating students' interdisciplinary competencies for problem-solving.

Calls for broad and interdisciplinary programmes prompted the re-emergence of liberal arts education in Europe (Van der Wende, 2011). As mentioned in the preface to this volume, an education characterised as interdisciplinary can naturally adopt a liberal arts educational philosophy. Hints of liberal arts in specialised educational programmes are already discernible, as evidenced by efforts to integrate the humanities and the social sciences into disciplines that have been traditionally monodisciplinary, such as engineering (Hitt et al., 2023) and medicine (Macleod & McCullough, 1994; Macnaughton, 2000; Schwartz et al., 2009). The broad character of a liberal arts education, which exposes students to various disciplinary knowledge, provides a fertile ground for nurturing their interdisciplinary ways of thinking.

Several European countries, including the Netherlands and Germany, have experienced a resurgence in the popularity of liberal arts education. Additionally, an increasing number of educational programmes in European universities have embraced interdisciplinarity as an educational approach. In the Netherlands, the University of Twente launched an interdisciplinary bachelor's programme in engineering in 2013, grounded in a liberal arts educational philosophy—the three-year bachelor's programme in Technology, Liberal Arts, and Sciences or ATLAS.

The third chapter by Wilhelm et al. describes the organisation and implementation of the aforementioned programme, predicated on its five core values: self-directed learning, interdisciplinarity, personal development, community, and trailblazing. These core values, the authors argue, are reflected in the formulation of the programme's final learning outcomes, which are anchored in the four learning lines—disciplinary lenses for understanding, processes of creation, communication and collaboration, academic approach and attitude—across the educational programme. The programme's emphasis on interdisciplinarity is a fundamental rationale for designing educational experiences that require students to pursue knowledge integration, which, according to Clark and Wallace (2015), is a principal feature of interdisciplinarity.

Additionally, the programme's adoption of problem-based learning (PBL) signifies that a real-world challenge must be at the centre of the learning process, as Kolmos (2009) proposes, especially in semesters where students must collaboratively complete a large project. Often considered a common approach in interdisciplinary engineering education (Van den Beemt et al., 2020), PBL requires a set of courses that can be related to the project's theme and scope (Kolmos, 2009) and can be used by students as lenses to understand and solve the real-world challenge embedded in the project. This specific requirement, underscoring the need for knowledge integration, is consequential for how mandatory courses within the semester are developed and organised.

Knowledge integration in a semester project demands a judicious selection of themes and concepts to be introduced in mandatory courses. This is one of the central propositions of the chapters that describe how courses in the social sciences (Chap. 5, Beldad), mathematics (Chap. 6, Kittou), and the natural sciences (Chap. 7, Van der Hoef et al.) are designed. Both Kittou and Van der Hoef et al. advocate for the importance of modelling—in mathematics and physics courses—to understand and solve real-world problems. Furthermore, according to Beldad, the social aspects of many real-world problems and of the solutions they require signify that interdisciplinary engineering students be introduced to the concepts, perspectives, and methods behavioural and scientists employ when studying behavioural and social phenomena. An important point that resonates throughout those three chapters is that an unconventional educational programme demands a novel approach to course development and constant experimentation with innovative didactic approaches.

Learning in PBL is primarily participant-directed (De Graaf & Kolmos, 2003), denoting that throughout the entire learning journey, learners must be placed in the driver's seat. Wilhelm thoroughly explores this proposition in Chap. 4. Throughout the three-year bachelor's programme, students hone their self-directed learning (SDL) skills, especially with the programme's open and broad character, which grants students considerable autonomy to design their educational curriculum.

At the start of a semester, as discussed in Chap. 4, students are required to write their personal development plans (PDPs) that state their learning needs and specify the educational activities for realising the broadly formulated semester learning goals. It is through this PDP that students take ownership of the learning process. In the middle of the semester, they evaluate how they are progressing in achieving the semester's learning goals according to their planned educational activities. For all students, the semester ends with an evaluation of their performance throughout their semester and an in-depth reflection on the implications of the results of their educational activities (e.g., courses) for their academic and personal growth.

Members of the academic staff guide students in planning their semester and act as discussion partners, particularly when students are evaluating their progress and performance and reflecting on their academic and personal development. Planning of educational activities, progress monitoring, performance evaluation, and reflection on personal growth constitute the SDL approach that the ATLAS programme adopts.

SDL permeates the entire bachelor's programme, especially as students have ample space to select their elective courses, develop new learning opportunities with experienced academics (e.g., capita selecta), and identify other avenues for learning beyond the classroom (e.g., internships). Another distinct feature of the ATLAS programme is the space students have to deepen their personal passion or interest within an academic environment, but beyond the confines of a standard curriculum. A Personal Pursuit, described in Chap. 8 by Dijksterhuis, is another learning experience that enables students to demonstrate their self-directedness. Through this 6-European credit Personal Pursuit, students can further develop their personal interests (e.g., learning a foreign language, documentary production, woodwork, nature photography) and acquire other valuable skills.

The five core values guiding the educational programme's vision of educating a new engineer are reflected in its various components. For instance, community is emphasised in the semester projects that students collaboratively work on. Interdisciplinarity manifests itself in the combination of courses students take within a semester. The value of trailblazing is fostered by consistently encouraging students to go beyond conventional expectations, explore diverse and novel routes to knowledge acquisition, and dare to pursue activities that may seem daunting. The promotion of these core values reflects the ATLAS programme's focus on nurturing students' academic and character development, which prompted the creation of the Academic Approach and Attitude (AAA) learning line.

As Kizito discussed in Chap. 9, the AAA learning line serves as a vehicle for students to develop metacognitive skills—such as analytical and critical thinking, logical reasoning—for intellectual development, as well as interpersonal, intrapersonal, and communal skills (e.g., cultural sensitivity) for character development. Various educational activities (e.g., courses, semester projects, the Personal Pursuit) afford students ample opportunities to develop those skills. For instance, assignments within a course are deliberately designed to sharpen their ability to think critically, while a semester project involving individuals from different sociocultural clusters stimulates them to be socially and culturally sensitive. Through this AAA learning line and the five core values, the programme clearly aims to educate future professionals who are not only intellectually competent but also professionally and socially responsible and ethically grounded.

Despite the expected benefits that could be derived from having a learning line that stresses both intellectual and character development throughout an interdisciplinary engineering programme, challenges to its implementation are inevitable. For one, there are difficulties associated with AAA-related educational activities throughout a semester. Objective evaluation of how students are realising the goals related to character development-related aspects (e.g., self-awareness and cultural sensitivity) also proved daunting and arduous. Then there is the persistent view among students of the peripheral value of AAA activities, considering them marginally relevant when compared to their course assignments and project deliverables. Nonetheless, according to Kizito, these challenges offer useful insights into how efforts to promote both academic and character development can be seamlessly integrated in an interdisciplinary engineering curriculum.

Empirical evidence for the positive effects of self-directed learning on students' growth throughout their learning process abounds. What remains less understood are the implications that years of experience with self-directed learning have for graduates' functioning in both professional and personal contexts. This is one gap in the current body of knowledge that the study by Kampouroglou et al. tried to fill. In Chap. 10, the authors discusses the results of in-depth interviews with 26 graduates of the ATLAS programmes across nine cohorts.

Interviewees regard SDL both as a way of thinking and an enabling tool for personal development. Across the 26 interviews, it became apparent that SDL significantly benefited graduates, as it enhances their sense of agency and autonomy, develops their self-reflection skills, strengthens their self-confidence, and sharpens their critical thinking skills. Nonetheless, experience with SDL is also associated with certain issues, such as graduates struggling to deal with failures and having an overly critical view of themselves and specific situations. Despite the perceived flaws in the implementation and operationalisation of SDL during their education, graduates' attitude toward SDL remains positive several years after graduation, primarily due to the many benefits they have acquired from an education that fosters SDL.

This edited volume opens with an incisive analysis of the societal value of inter- and transdisciplinary education and the liberal arts educational programme by Boor et al. and Kumar et al., respectively. The succeeding chapters describe how an interdisciplinary engineering programme founded on a liberal arts educational philosophy is organised and implemented, taking into account its various components—from the educational programme's core values and final learning outcomes to its various educational activities (e.g., mandatory courses, projects). A common theme coursing through the different chapters is the instrumental role of an interdisciplinary engineering programme in forming well-rounded engineering professionals - intellectually competent and socially oriented to help mitigate the challenges of the twenty-first century.

Nonetheless, as Chap. 11 by Beldad et al. reveals, organising and implementing such an educational programme is far from being a problem-free enterprise. The open and broad character of a liberal arts education, the freedom to personalise and to choose afforded by self-directed learning, and the requirements of having an interdisciplinary community of learners and educators are catalysts for the challenges confronting programme implementation. Throughout this final chapter, the authors analyse the curricular (e.g., tension between maintaining interdisciplinarity and the urge to specialise in a student's individual curriculum) and organisational hurdles (e.g., interpersonal dynamics within a multidisciplinary team of academics) an innovative educational programme faces. The approaches the programme took to counter those challenges are also described.

Clearly, designing an interdisciplinary education has its complexities, while a liberal arts education has its share of issues.Undoubtedly, monodisciplinary education still has its merits. However, there is no denying that the many complex problems we all face compel innovative educational approaches to train a new breed of engineers possessing the competencies and the character needed for interdisciplinary and socially responsible problem-solving. What the world increasingly needs are problem solvers adept at melding knowledge from multiple disciplines when navigating complex challenges. Organising and implementing an interdisciplinary engineering education with a strong liberal arts character may not be plain sailing, but the societal significance of such an education can and must never be dismissed.

References

Clark, S. G., & Wallace, R. L. (2015). Integration and interdisciplinarity: Concepts, frameworks, and education. *Policy Sciences, 48*(2), 233–255. https://doi.org/10.1007/s11077-015-9210-4

De Graaf, E., & Kolmos, A. (2003). Characteristics of problem-based learning. *International Journal of Engineering Education, 19*(5), 657–662.

Hitt, S. J., Banzaert, A., & Pierrakos, O. (2023). Educating the whole engineer by integrating engineering and the liberal arts. In A. Johri (Ed.), *International handbook of engineering education research* (pp. 457–476). Taylor & Francis.

Kolmos, A. (2009). Problem-based and project-based learning. In O. Skovsmose, P. Valero, & O. R. Christensen (Eds.), *University science and mathematics education in transition* (pp. 261–280). Springer.

MacLeod, S. M., & McCullough, H. N. (1994). Social science education as a component of medical training. *Social Science & Medicine, 39*(9), 1367–1373. https://doi.org/10.1016/0277-9536(94)90367-0

Macnaughton, J. (2000). The humanities in medical education: Context, outcomes and structures. *Medical Humanities, 26*(1), 23–30. https://doi.org/10.1136/mh.26.1.23

Rhoten, D., Boix Mansilla, V., Chun, M., & Klein, J. T. (n.d.). *Interdisciplinary education at liberal arts institutions*. Teagle Foundation White Paper, 13.

Schwartz, A. W., Abramson, J. S., Wojnowich, I., Accordino, R., Ronan, E. J., & Rifkin, M. R. (2009). Evaluating the impact of the humanities in medical education. *Mount Sinai Journal of Medicine: A Journal of Translational and Personalized Medicine, 76*(4), 372–380. https://doi.org/10.1002/msj.20126

Van den Beemt, A., MacLeod, M., Van der Veen, J., Van de Ven, A., Van Baalen, S., Klaassen, R., & Boon, M. (2020). Interdisciplinary engineering education: A review of vision, teaching, and support. *Journal of Engineering Education, 109*(3), 508–555. https://doi.org/10.1002/jee.20347

Van der Wende, M. (2011). The emergence of liberal arts and sciences education in Europe: A comparative perspective. *Higher Education Policy, 24*(2), 233–253. https://doi.org/10.1057/hep.2011.3

Glossary

ATLAS The three-year bachelor's programme Technology and Liberal Arts & Sciences that emphasises the equal roles of the natural sciences and the social sciences in generating solutions to the complex socio-technical challenges of modern society. The programme was inspired by the idea that new technologies are crossing traditional disciplinary boundaries, which implies that engineering solutions require an integrated perspective, especially in the way technologies are introduced into and affect society.

Challenge-based Learning (CBL) A form of inquiry-based learning that provides students with opportunities to engage in open-ended real-world challenges, requiring the adoption of multidisciplinary perspectives, self-directed collaborative work, and critical and creative thinking.

European Credit (EC) European credits are units in the European Credit Transfer and Accumulation System (ECTS) used across Europe to measure student workload and make academic qualifications comparable, facilitating study abroad and credit transfers. One EC typically equals 28 hours of study, with a full academic year equivalent to 60 European credits (around 1,680 hours).

Final Learning Outcomes (FLOs) FLOs refer to a of goals that every student has attained after completing the educational programme.

Inner Development Goal (IDG) Framework A framework that inspires embedding affective dimensions (attitudes, beliefs, emotional capacities) in the curriculum, by outlining skills, attitudes, and values across five dimensions (being, thinking, relating, collaborating, acting).

Inter- and Transdisciplinary (ITD) Education Education that focuses on integrating methods, concepts, and theories from disciplines to generate new (interdisciplinary) understandings and on incorporating extra-academic sources for knowledge, such as entrepreneurs, policymakers, community members, patient organisations, and practitioners, to produce an expanded (transdisciplinary) scope.

A. Beldad et al. (eds.), *Designing an Interdisciplinary Engineering Education for Complex Real-World Challenges*, Emergent Directions in Engineering Education 1, https://doi.org/10.1007/978-3-032-20223-9

Learning Line A series of educational activities that build upon each other and serve a shared set of final learning outcomes.

Liberal Arts & Sciences Education An international type of education characterised by exposing students to a wide range of academic subjects and the use of a personal, student-centred approach.

Mid-Term Evaluation (MTE) In the ATLAS programme, the MTE is performed at the midpoint of a semester and aims to evaluate the student's progress towards achieving the planned semester goals. This provides students with the opportunity to make modifications in their study plans to ensure the achievement of those goals.

Problem-Based Learning or Project-Based Learning (PBL) A form of inquiry-based learning that primarily involves students constructing an end product (a 'concrete artefact') that displays both knowledge and skills acquired from courses and their ability to apply knowledge to solve a real-world problem. Some experts use problem-based learning when the real-world problem is more predefined, and the answer is more known, and use project-based learning when the real-world problem and the solutions are unclear and unknown.

Personal Development Plan (PDP) In the ATLAS programme, students create an individual study plan (the PDP) for a semester detailing how and with what evidence they intend to collect to demonstrate the attainment of a semester's learning goals.

Personal Pursuit (PP) An educational activity (part of the ATLAS programme) in which a student transforms a personal interest into an academic educational project. Students design their own learning by formulating learning outcomes, specifying learning activities, and the final product of the project.

Self-Directed Learning (SDL) A form of education that offers students autonomy and choice in designing their educational curriculum. Students personalise learning goals to match their self-identified learning needs, find ways to satisfy those needs, and decide on how to determine successful learning goal attainment.

Self-evaluation and Reflection (SER) In the ATLAS programme, an evaluation (the SER) is performed at the end of the semester. In the form of a presentation, the student substantiates their learning with evidence and reflects on their academic development.

MIX
Papier aus verantwortungsvollen Quellen
Paper from responsible sources
FSC® C105338

If you have any concerns about our products,
you can contact us on
ProductSafety@springernature.com

In case Publisher is established outside the EU,
the EU authorized representative is:
Springer Nature Customer Service Center GmbH
Europaplatz 3, 69115 Heidelberg, Germany

Printed by Libri Plureos GmbH
in Hamburg, Germany